C0-AVC-618

Adaptations of Desert Organisms

Edited by J.L. Cloudsley-Thompson

Springer
Berlin
Heidelberg
New York
Barcelona
Budapest
Hong Kong
London
Milan
Paris
Tokyo

Volumes already published

Ecophysiology of the Camelidae and Desert Ruminants
By R.T. Wilson (1989)

Ecophysiology of Desert Arthropods and Reptiles
By J.L. Cloudsley-Thompson (1991)

Plant Nutrients in Desert Environments
By A. Day and K. Ludeke (1993)

Seed Germination in Desert Plants
By Y. Gutterman (1993)

Behavioural Adaptations of Desert Animals
By G. Costa (1995)

In preparation

Invertebrates in Hot and Cold Arid Environments
By L. Sømme (1995)

Bioenergetics of Desert Invertebrates
By H. Heatwole (1995)

Giovanni Costa

Behavioural Adaptations of Desert Animals

With 56 Figures

Prof. Dr. Giovanni Costa
Università di Catania
Dipartimento di Biologia Animale
Via Androne 81
95124 Catania, Italy

Cover illustration: photograph by J.L. Cloudsley-Thompson

ISBN 3-540-58578-8 Springer-Verlag Berlin Heidelberg New York

Library of Congress Cataloging-in-Publication Data. Costa, Giovanni, 1942– . Behavioural adaptations of desert animals/Giovanni Costa. p. cm. – (Adaptations of desert organisms) Includes bibliographical references (p.) and index. ISBN 3-540-58578-8 (Berlin). – ISBN 0-387-58578-8 (New York) 1. Desert fauna – Behavior. 2. Adaptation (Biology) I. Title. II. Title: Behavioral adaptations of desert animals. III. Series. QL 116.C65 1995 591.5'2652 – dc20 94-41930

Printed in Germany

Typesetting: Macmillan India Ltd., Bangalore-25

SPIN: 10122749 31/3130/SPS – 4 3 2 1 0 – Printed on acid-free paper

Preface

From the beginning of my scientific activity, I have been engrossed with the study of animal behaviour. I have always been convinced that animals, as regular inhabitants of specific habitats, evolved winning strategies to adapt to apparently adverse environmental conditions (and more generally to their ecosystems); so it has been natural for me to carry out ecoethological research. I began by studying animals living in sandy/coastal ecosystems and, subsequently, took a great interest in desert animals. In this environment, I expected to find striking adaptations but I must admit that even my fertile imagination was surpassed by what I learnt through both personal experience and the study of the rich scientific literature on the behaviour of desert animals. For this reason I accepted – with pleasure and also with some concern – the invitation of my friend John Cloudsley-Thompson to contribute, with this work of mine, to the series *Adaptations of Desert Organisms*. I hope that this book will be useful to its readers and will discredit some myths based on incorrect but widespread beliefs, concerning deserts and their faunas.

Indeed, many people believe that the desert is a lifeless environment – in everyday language 'desert' is almost synonymous with 'uninhabited'. On the contrary, however, the student of the biology of arid environments knows very well that they are populated by a rich and varied fauna. In connection with this first error, is another, according to which the desert is a hostile habitat: but, since it is true that a desert biotope presents unbearable life conditions for organisms from other ecosystems, the desert promotes its own biocoenosis: in fact, the latter has survived there for a very long time.

However, what I have said is true for every environment: almost any organism, if moved from its natural habitat to another, will respond like a "fish out of water". It must be borne in mind that the more selective a biotope is, the more specialized its biocoenosis becomes: plants and animals are strictly linked with their habitat, and their particular adaptations are the result of this relationship. This, together with the decisive role of behaviour in animal survival, is what justifies this book.

Catania, Spring 1995 G. Costa

Acknowledgements

I wish to express my gratitude to the editor of this Series, Prof. John Leonard Cloudsley-Thompson. He provided me with advice, suggestions and help. In addition, he undertook the hard task of revising the text and translating my 'dog' English into intelligible language.

My thanks are also due to Dr. Dieter Czeschlik, Biology Editorial of Springer-Verlag, for his confidence and boundless patience.

I am deeply grateful to Dr. Erminia Conti, my pupil, who assisted me in typing the manuscript and preparing the original illustrations. Moreover, the responsibility for encouraging me to continue working, whenever I wanted to abandon the undertaking rests entirely with her!

Contents

1 Introduction

It is an old concept that, since life first appeared on Earth, it has undergone continuous transformation which is responsible for both the great variety of living beings and the wonderful correspondence between organisms and their environments. In his book, *The Origin of Species* (1859), Darwin formulated the theory of evolution by natural selection. According to this theory, the 2 million species present today developed from pre-existing species. (It is estimated that, since the beginning of the Cambrian period, about 600 million years ago, about 2000 million species have appeared on Earth.) Furthermore, Darwin realized the fundamental role of the congruence of every living being with its own habitat. The environment presents many problems, changeable in the long run, that organisms must solve if they are to survive: species incapable of finding suitable solutions are destined to die out; in contrast, those that are able to face unforeseeable changes in the environment can keep themselves alive and leave their offspring endowed with the same potentialities.

Recent scientific, in particular genetic, discoveries have totally confirmed the Darwinian theory of natural selection which, even if more than once contested and opposed, still provides the most satisfactory explanation of the organization and development of life on Earth.

The evolutionary changes, through which organisms, little by little, have developed better solutions to environmental demands, are called 'adaptations'. They are the result of various progressive modifications, both structural and functional. Success in the continuous struggle for survival depends on the interaction of these modifications. In the case of animals, the structural and functional levels are integrated with a third one, the behavioural level, which ends by becoming fundamental. In fact, some constitutional characteristics of animals, such as mobility, extreme sensitivity, and heterotrophism, create the bases for a dialectic interrelation between each of these organisms and its habitat. According to natural selection, the living individual will become 'adapted' to the environment, and so will become better at finding food or sexual partners, and at avoiding climatic extremes or predators.

In most cases, it is evident that the adjustment of animal behaviour is based on ascertainable morphological and functional features. For example, the physical characteristics of vocalizing or singing in species provided with acoustic communication depend strictly on the structure of the sound-producing organs, and on the neurohormonal control of sound production. Likewise, the ability of some animals to dig burrows in the ground is closely linked with fossorial structures and their neuromotor co-ordination. Similarly, animal camouflage means utilizing a particular body shape and colouring; and so on. This interaction may make the analysis of every ethological adaptation very difficult, but it is beyond doubt that behaviour involves the most elevated level of adjustment in the animal kingdom. So, morphological and functional modifications would have no efficacy, if they did not combine with definite patterns, such as those correlated with singing, burrowing or the appropriate spatial orientation of the body.

In the last few decades the concept of inclusive fitness has been affirmed: it is the role of an individual to contribute to the genetic pool of the population – either directly, through its own reproductive success, or indirectly, through so-called altruistic behaviours. Thus, the individual increases the survival of its relatives with which it has several common genes (Hamilton 1964; Dawkins 1978). According to sociobiologists, 'animal altruism', once considered a biological paradox, has assumed the meaning of 'genetic egoism'. For example, the famous 'sacrifice' of worker bees towards their sisters is now considered the epitome of efficient behaviour, allowing sterile individuals to perpetuate the maximum quantity of their genes. As Hymenoptera are haplodiploid animals, a female has more genes in common with her own sisters than she has with her daughters. In fact, each female receives half her genes from her haploid father and the other half from her diploid mother; so each female receives all the paternal and half of the maternal genes. Consequently, mother and daughter have only 50% of the hereditary patrimony in common (coefficient of relation 1/2), whereas two sisters will have all the paternal genes and half the maternal genes in common (coefficient of relation 3/4). This simple but substantial difference between the two coefficients of relationship explains why a particular social organization among bees and also among other haplodiploid insects has evolved.

The advantages of an ethological pattern do not always depend on genetic adaptations of the behavioural inventory (ethogram) of the species. In many cases, they may be the result of accidental effects and, therefore, are not due to natural selection. During ontogeny, an animal may be in a situation compelling it to behave in a more or less different way than usual: a fox can recognize its own tracks in the snow after some time, if these are still visible, so it can quickly reach a henhouse previously visited. A Japanese macaque, which usually eats muddy potatoes, may suddenly discover the cleansing effect of water after a potato has slipped and fallen into a brook (Kawamura 1954; Kawai 1965). An English blue-tit may, by mere chance, discover an excellent new food, by piercing the cap of a milk bottle with its bill to reach the cream concentrated at the top (Fisher and Hinde 1949; Hinde and Fisher 1952). An ant, put into a labyrinth, is able, after unsuccessful attempts, to find a way out (Schneirla 1953); and so on. These occasional experiences would be devoid of positive consequences, if the animals had not inherited – a fundamental genetic adaptation – the bent for memorizing them, and therefore for learning new advantageous patterns. For this reason an animal, which has made a discovery, is apt to repeat and 'imitate' itself to obtain the same advantage again. Many species with a complex social organization are able to transfer to other members of their own group what one of them has learned. This is what happened in the case of the Japanese macaques and English tits.

Thus, 'cultural' evolution is added to genetic evolution. It allows many different behaviours only if the animals have a genetic tendency for practice and imitation.

There are many examples proving the determining influence of peculiar biotopic characteristics on animal behaviour. Cloudsley-Thompson (1975) clearly explained the role of climate and the topographic peculiarity of terrestrial environments in the animal kingdom. In forest ecosystems, many species are adapted to arboreal life, with the help of structures such as strong arms (primates such as chimpanzees and monkeys), prehensile feet (many monkeys) or tails (chameleons), crooked claws (North American tree porcupines and sloths), adhesive pads or discs (arboreal frogs and geckoes), alar membranes (the extinct pterodactyls and existing flying amphibians, mammals and reptiles), and so on. Convenient behavioural patterns and body features concur to make the congruence of arboreal animals with their habitat more satisfactory. In open country, speedy locomotion and excellent endurance, helped by lengthening of the legs, have evolved (Gray 1968). (Ostriches, bustards, horses and antelopes are typical representatives of cursorial animals.) In many species, burrowing habits, helped conversely by shortening and strengthening of the legs, have evolved (Lull 1940).

In aquatic environments, there are several adaptations that make survival possible. These differ, of course, according to how recently colonization of the water habitat has occurred.

Desert is certainly one of the most selective of terrestrial biotopes for both plants and animals: in fact, these show similar morphological and physiological adaptations which permit them to face the severity of the climate and to minimize water loss (Hadley 1972).

Moreover, in regard to behaviour, desert animals provide excellent examples of both autoprotective (thermohygric regulation, movement patterns, antipredatory strategies) and alimentary, reproductive and social adaptations. In particular, animals show the most striking adaptation to environmental problems in sandy areas. All these subjects will be discussed in the following chapters.

2 Biotope and Vegetation Features

2.1 Definition of Desert

From a study of the surface of our planet, about one third of all the land consists of territories with distinctively arid features (Fig. 1). Deserts form a very heterogeneous whole due to the differing climatological and geological histories of each of them. Scientists agree that aridity is the common denominator. Nevertheless, there is some disagreement about the way in which temperature, scarcity of rainfall, evaporation, etc. interact in causing desert conditions. Moreover, the criteria used to grade various arid regions of the Earth and to determine their limits are very subjective.

In considering the severity of arid conditions, Koeppen and Geiger (1930) divided the terrestrial environments into desert, semi-desert and moist areas. Accordingly, the boundary between moist and semi-desert environments lies where potential evaporation equals precipitation, while the boundary between desert and semi-desert environments lies where evaporation is two-fold that of precipitation.

On the basis of rainfall only, McGinnies (1968) divided arid areas into: hyperarid (less than 100 mm mean annual precipitation), arid (from 100 to 250 mm) and semi-arid areas (from 250 to 500 mm). Noy-Meir (1973) pointed out that, more than annual rainfall, irregularity and unpredictability of precipitation are the main factors creating arid conditions.

More generally, it is not so much the quantity of water present as the quantity of water readily exploitable by plants that determines desert characteristics. In arid environments, primary productivity is low (less than 0.1 g mm^{-2} day^{-1} in comparison with over the 10 g mm^{-2} day^{-1} in forest ecosystems). These values are comparable only with those of the deep sea, where light is the limiting factor.

In support of this ecological definition of desert, we should remember the 26 million km^2 of polar areas where, independently of precipitation, a great quantity of water is present throughout the year in the form of ice, which plants cannot utilize. Moreover, there are several zones which remain arid even though plentiful rains may occur. This is due either to the great speed of evaporation caused by high temperatures, or to soil porosity – as in sandy-coastal environments where rainwater filters through the substratum too rapidly for plants to gain advantage from it.

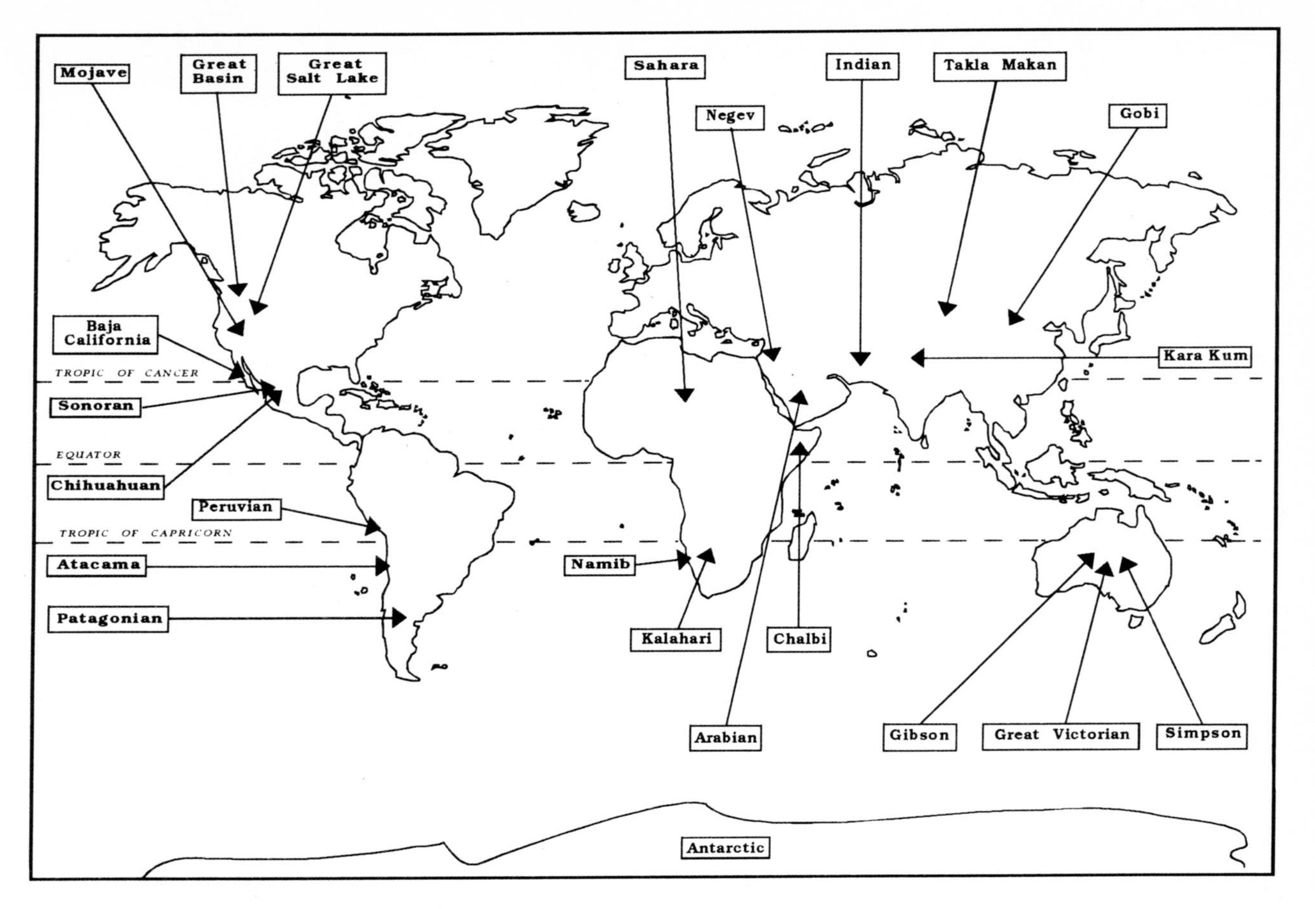

Fig. 1. Map showing the world's major deserts

2.2 Factors of Aridity

During the geological and climatological history of the Earth, conditions of humidity and aridity have arisen many times and over large areas. Present deserts are relatively recent formations – a few millions of years old according to palaeobotanical estimates or a few tens of millions of years according to geological estimates. Animal and vegetable species, adapted to arid environments, evolved during this time.

Various climatic, geomorphological and geographical factors cause desert conditions. A 'superfactor' is the rotation of the Earth around its axis (rotation velocity is about 1667 km h^{-1} at the equator and zero at the poles), which causes a particular air circulation and a regular alternation of low and high pressure zones from the equator to the poles. It also gives rise to oceanic currents, which contribute strongly to determine the aridity of extensive coastal areas.

Other important factors may be the distance of large continental regions from an oceanic source of humidity or the existence of large mountain barriers, which cause the so-called rain shadow effect.

2.3 Classifications of Deserts

A traditional classification of the arid zones of the world, based on climatic characteristics, recognizes five principal types of desert: (1) tropical and subtropical deserts; (2) cool coastal deserts; (3) rain shadow deserts; (4) interior continental deserts; and (5) polar deserts. A more complete survey should also include: (6) sandy-coastal environments and (7) semi-desert areas.

2.3.1 Tropical and Subtropical Deserts

Between approximately 30°N (Tropic of Cancer) and 30°S (Tropic of Capricorn) there are two distinct desert belts separated by a large circumequatorial band of forests and savannahs. They are the result of worldwide circulation patterns of the air which, at the level of the tropics, produce semi-permanent cells of high pressure within which the air has a tendency to descend from high altitudes towards the Earth's surface. When this very dry and initially cold air reaches the ground, it is very hot due to the compression to which it is subjected. (The adiabatic rate of warming is about 10 °C for 1000 m.) These characteristics of the air cause the aridity of tropical and subtropical deserts. The tropical arid belt of the Northern Hemisphere includes, among the more representative deserts, the Sahara of North Africa (Fig. 2), the

Fig. 2. Sandy area of the Sahara Desert showing a herd of *Camelus dromedarius* in a semi-natural state

Arabian and Oriental deserts of the Old World and the Sonoran Desert of the New World. The tropical arid belt of the Southern Hemisphere is represented by the Kalahari Desert of southern Africa, by a great many of the Australian deserts, and by small areas of western Argentina.

2.3.2 Cool Coastal Deserts

It may seem paradoxical that lands lapped by an ocean may be desert; nevertheless, there are at least three large coastal areas with desert characteristics: the Namib along the Atlantic African coast (Fig. 3), the Atacama and Baja California deserts along the Pacific coast of South and North America respectively.

In these deserts, descending high pressure air masses, responsible for the lack of rain, add to the influence of cold oceanic currents coming from the polar regions: the Benguela Current (Namib), the Humboldt Current (Atacama) and the Californian Current (Baja California). The coastal zones are made cold and foggy by the cold water brought by these currents.

Fig. 3. Namibian gravel plain at Gobabeb. In the background, the linear oasis of the Kuiseb River bounds the impressive dune system

2.3.3 Rain Shadow Deserts

These are areas close to the lee side of large mountain ranges. Damp air, sometimes oceanic, colliding with these natural barriers, relieves itself of moisture on the windward side giving rise to abundant precipitation (orographic precipitation), and arrives at the opposite side quite dry, causing the aridity of the lee areas. The deserts belonging to this group are: the Mojave Desert in California, bordered by the Sierra Nevada and the Transverse Ranges, the Great Basin Desert in Nevada, Utah and adjacent states, sheltered by the Sierra and Cascade Ranges to the west and by the Rocky Mountains to the east, and the Patagonian Desert in Argentina, where the climatic situation is influenced by the orographic system of the Andes to the west and by South Atlantic cold air masses to the east. Some Australian arid zones, including those close to the Great Dividing Range, present the same characteristics as rain shadow deserts.

2.3.4 *Interior Continental Deserts*

Many lands, situated great distances from the sea, may be desert owing to their continentality. The most typical case is that concerning the inland Asiatic areas; however, many examples are found in Australia, North America and in large areas of the Great Palearctic Desert. Lack of the moderating influence of the sea causes severe climatic conditions (extremely cold winters and extremely hot summers). Furthermore, precipitation is very low, because winds arrive without moisture, which has been lost on the way from the sea.

2.3.5 *Polar Deserts*

Extremely low temperatures prevalent throughout the year cause the aridity of polar areas (essentially, Greenland and the Arctic coast of Canada, Alaska and Siberia in the Boreal Hemisphere and Antarctica in the Austral Hemisphere). Evaporation is low: moisture condenses into ice spicules floating in the air; slight precipitation falls as snow (not above 130 mm year^{-1}), and water accumulates in glaciers. These climatic features are similar to those

Fig. 4. Southern part of Tierra del Fuego, north of Antarctica, during the Austral summer

existing at high altitudes on mountains at all latitudes. Plants have no available water, which is present only as ice, making all these areas 'physiological deserts' (Fig. 4).

2.3.6 Sandy/Coastal Areas

Incoherence and the porosity of sandy soil enhance the rapid percolation of rainwater, and cause the extreme aridity of the superficial soil layers of sandy/coastal areas. The intense solar radiation quickly dispels the humidity caused by night dew and contributes to aridity (McCallan 1964). Considerable thermal fluctuation and the influence of the sea complete the principal features of this 'edaphic desert' (Fig. 5). In this environment, there is a definite zonation, with clear sandy belts (differing in biotopic and vegetational characteristics) which follow each other from the water's edge to the inner dune areas.

2.3.7 Semi-Desert Areas

Regions of transition between arid and damp areas are generally situated at the edge of semi-permanent, high-pressure areas or in the rain shadow of a

Fig. 5. Atlantic sandy beach, inhabited by various bird species, such as pelicans and cormorants

continental cordillera. From a vegetational point of view, semi-desert areas are grasslands – that is, environments distinguished by the dominating grasses and by the acute scarcity of arboreal and shrubby plants. Temperate grasslands are usually called ‘steppes’. These are prevalently located in central Eurasia, but they are also found in North America and in the Southern Hemisphere. In the such places, they have local names (‘pampas’ in Argentina, ‘veld’ in South Africa or simply ‘prairies’ in North America and Australia). A further step towards a closed vegetational environment is presented by savannah, where there are considerable number of shrubs and the yearly precipitation can amount to almost 1000 mm.

A particular case of a semi-desert area is the ‘tundra’ (or cold prairie) which can appear in two forms: Arctic tundra and Alpine tundra. The first, situated between the polar Arctic desert and the taiga (cold coniferous forest of subarctic latitudes of the Northern Hemisphere), is characterized by long cold winters and short cool summers during which the superficial soil stratum thaws while the subsoil remains permanently frozen as permafrost, as in the polar desert. Annual precipitation, not greater than 350 mm, falls during the warmest period of the year. Water which is not absorbed by the superficial layers of the soil accumulates in the form of extensive swamps. According to several authors, the Arctic tundra, in spite of the phenomenon of summer thaw, can be included among the polar Arctic deserts (in the Southern Hemisphere, considerable ice thickness makes the formation of tundra impossible).

The Alpine tundra found at southern latitudes covers mountain slopes. Precipitation is more frequent here, but the marshy and rocky ground is similar to that of Arctic tundra.

2.4 Biotope Features

Biotope features are of paramount importance for the peculiar adaptations of animals to the environment. Biotope, indeed, is both the scenarist and the director of the tangled and long history of the evolution of life on Earth. The desert has not evaded this double role. Indeed, taking the severity of the structural and climatic features into account, it has probably exerted, and exerts, the greatest moulding force on its inhabitants. So, it may be useful to examine, even if briefly, some its parameters before discussing the behaviour of the fauna.

2.4.1 Temperature

Temperature is a fundamental climatic factor. Many arid and semi-arid regions are characterized by very high temperatures. With regular seasonal cadence, a thermal maximum is reached at midday because solar radiation is not reduced by clouds or atmospheric humidity. Consequently, heat emission from the overheated ground is also not reduced. Direct solar radiation has a second important biological effect: cellular damage may be caused by direct UV penetration of tissue. Most desert organisms have evolved protection against exposure to UV (Larmuth 1984). Exceptional thermal peaks are recorded every so often: for instance, 58 °C in the Sahara, 56.5 °C in the Great Basin, and 51 °C in the Sonoran desert. More frequently, long periods with a mean maximum above 40 °C are recorded. Consequently, the surface temperature of the ground, especially when the substratum is sandy, can reach values between 60 and 70 °C at midday, giving rise to serious problems for the survival of animals in an environment without vegetation cover. Over 5 days in September 1962, Cloudsley-Thompson (1964) recorded 84 °C on Nile silt at Wadi Halfa (Sudan).

A further important factor is the diurnal thermal range. Because of the strong nocturnal fall in temperature (reaching values below zero in the coldest season), this may reach 40 °C. The annual range of shade temperature may sometimes exceed 55 °C.

Of course, temperature extremes and thermal fluctuations are gradually reduced in the lower layers of the ground. In Saharan dry sandy soils, the thermal change is about 5–10°C at a depth of 10 cm and only of about 1°C at 50 cm, during daytime extremes at the surface (Larmuth 1984). At night, while the surface cools, the deeper layers remain warm.

On the basis of the mean monthly temperature (mmt) of the warmest month, deserts can be labelled *hot* (mmt > 30 °C), *warm* (mmt = 22–30 °C), *cool* and *cold* (mmt = 10–22 °C), and *polar* (mmt < 10 °C). The difference between cool and cold deserts is based on the mean monthly temperature of the coldest month, which averages between 10–22 °C in the former, below 10 °C in the latter. Typical hot deserts include the interior continental deserts and are situated in tropical and subtropical areas and in the North American and Australian rain shadow. Warm deserts include some coastal zones such as the Mediterranean edges of the Sahara, the southern end of Baja California, and the southwestern coast of Madagascar. Cool coastal deserts were described previously. True cold deserts are the Central Asian and the Patagonian deserts. In the cold deserts, tundra and the polar deserts, where the temperature often drops far below zero, plants and animals must minimize the risk of frost damage. The minimum temperature can reach −50 °C in the Central Asian desert and −68 °C in Greenland. The maximum annual change in shade temperature in any terrestrial environment is to be found in tundra, where the winter minimum may reach −57 °C and the summer maximum 21 °C, with a thermal fluctuation very close to 80 °C!

2.4.2 Atmospheric Humidity

Very scarce and erratic rainfall causes the general dryness of arid environments. In addition, solar radiation, unopposed by plant cover, results in the rapid evaporation of rainwater (Louw and Seely 1982). This greatly reduces the effectiveness of rainfall in increasing ground moisture, but provides a sporadic contribution to the atmospheric humidity. The latter depends fundamentally on local climatic features and, to a certain extent, on the evapotranspiration mechanisms of the biocoenosis. This leads to many possible adaptations to minimize water loss.

Coastal desert areas can easily experience a relative humidity as high as 100%. In the inner regions, such values are attained only near oases or other stretches of water. In general, however, relative humidity seldom exceeds 50%. Some exceptions have been recorded. For example, the daily cycle of relative humidity in the Sudan during December 1961 ranged between 40% at midday and 85% at night (Cloudsley-Thompson 1966). Typical hot coastal deserts with oppressively high humidity include the southern coast of the Red Sea (tropical desert) and the coasts of the Persian Gulf (subtropical desert), where the high water and air temperature engenders strong evaporation from the sea. In the cool coastal deserts, the cooling effect of cold water offshore brings about the formation of a fog blanket which drifts onto the adjacent land. Fog appears, often early in the morning, in the Atacama in winter, and in Baja California and the Namib in summer. Fog and the consequently reduced visibility greatly influence the biology of many desert animals (Fig. 6).

Fig. 6. Gemsbok (*Oryx gazella*) moving under the damp morning fog

In the polar deserts, the relative humidity is also quite high, often reaching 85%; however, the low temperature induces the condensation of ice spicules which float in the air, which greatly limit visibility.

2.4.3 Wind

Wind is another very important climatic element of desert environments, and greatly contributes to their aridity. Moreover, wind speed is a determining factor for both the rate of convective heat transfer between the surrounding air and the bodies of animals, and the rate of evaporative water loss by all the desert inhabitants (Mitchell 1977).

Wind is also a significant agent regarding the surface modelling of the Earth in arid environments, where even the smallest sediments do not cohere due to the lack of moisture and vegetation. Strong winds, sometimes reaching a speed of 80 km h^{-1}, produce dust storms and sandstorms which erode the rocky ground and, in the long run, level it. Running water, both infrequent and extremely vehement in the desert, concurs with the wind in this process. The older a desert, the flatter it is.

Sand particles, shifted by wind, move across the ground blunting all the obstacles and accumulating to form dune systems. These develop along the coastal strips of seas and lakes, on the lee sides of rivers and in many desert territories. Here, large quantities of sand particles, arising from the disintegration of granite, sandstone, gypsum, etc., crowd in distinctive formations on the ground according to the direction and speed of the prevailing winds. Moderate winds with a uniform direction cause the formation of barchan dunes when sand is scarce, or of transverse dunes when sand is more abundant. Unidirectional and stronger winds give rise to longitudinal dunes. Pluridirectional winds are responsible for the formation of stellar dunes. Due to wind action, large desert zones may remain sandless, exposing a typical mosaic of gravel – as in the Saharan 'reg' – or of pebbles as in the northeastern African 'serir'. (This is the converse of 'erg', an area with a large quantity of sand.) Fragmentation of rock depends on rapid and wide thermal fluctuations or on the hydration and expansion of the constituent mineral grains caused by moisture. Rare rainstorms combine to modify the bare rock and the desert landscape (Cloudsley-Thompson 1977a).

2.5 Characteristics of Desert Vegetation

As in every ecosystem, the existence of a rich and varied zoocoenosis in the desert is based on the vegetation. The latter may seem fragile, but is actually endowed with almost unbelievable resistence. It results from the hyperspecialization achieved during the colonization of the particularly harsh environment where drought, except for episodic events, reigns without opposition.

Only in oases and along rivers is the vegetation thick. In arid environments, plant distribution is characteristically thin. Vegetational stratification, typical of forest ecosystems, is absent. It is not necessary to maximize the exploitation of sunlight. On the other hand, water is the most important limiting factor. Thus, there is reduced spatial competition among individual plants. However, there is considerable evidence of the underground struggle of plant roots to monopolize the scarce water resources at their disposal. The struggle is based, moreover, on the emission of toxic substances, which prevent new buds from sprouting. Likewise, several shrubby plants, such as the North American *Encelia farinosa* and *Thamnosma montana*, are known, whose leaves secrete poisonous chemical compounds, inhibiting the growth of other plants (Gray and Bonner 1948; Bennet and Bonner 1953).

On the basis of their water requirements, various classifications of arid zone plants have been proposed. A basic distinction separates plants which inhibit the negative effects of drought, and plants which face drought with suitable morphofunctional devices. A complete classification was suggested by Cloudsley-Thompson (1975), who distinguished four different situations.

a) *Drought-escaping species*: these are ephemeral plants, which can remain as seeds for many years; when the first rain comes, these seeds are able to sprout immediately. Rapid flowering abruptly changes the melancholy face of the desert landscape to a multicoloured earthly paradise and, in a few weeks, the whole life cycle of these extraordinary plants is completed. The seeds produced will wait patiently for the next rain.

b) *Drought-evading species*: because of their tiny size, these plants are able to retain the small quantity of water necessary for their limited growth.

c) *Drought-enduring species*: like the American 'creosote bush', *Larrea tridentata*, *L. divaricata* and *L. cuneifolia*, these plants are able to shed leaves, twigs and whole branches when soil moisture is absent. They may appear to be dead, however, when rain comes, small leaves and buds are produced by the branches and the plants start to grow again.

d) *Drought-resisting species*: these are succulents (Fig. 7), capable of holding large reserves of water in their tissues and storage organs (leaves, stems, underground apparatus).

As in all classifications, this is a simplification, and other examples can be included in more than one of the above-mentioned categories. For instance, several desert plants can belong both to c) and to d): they are not only able to remain vital while retaining only the underground structures, but they can also accumulate both water and chemicals for future use. Among the so-called root-succulent plants, there is a Kalahari wild vine (*Raphionacme burkei*, named 'bi' by Bushmen), whose tuber, full of water and as large as a basketball, is eagerly sought by thirsty Bushmen.

Fig. 7. The succulent *Euphorbia damarana* is common in the northern areas of Namibia

Lower plants have not been adequately studied in desert areas; little is known about their taxonomy and biology. More attention has been devoted to the vascular plants. The aboveground portion of these plants is often a herb or shrub, more rarely a tree. The herbaceous plants, widely dispersed in semi-arid areas, are generally showy post-rainfall ephemerals. They are sometimes the most distinctive of desert plants. Because of the prevalence of grasses, the drier regions of Australia can be called 'grass deserts' (McCleary 1968).

The vegetation of various deserts is dominated by particular shrubs and trees. Thus, the American deserts are characterized by large cacti, the Namib by the millenary dwarf gymnosperm *Welwitschia mirabilis* (Fig. 8), and the Kalahari by the huge baobab *Adansonia digitata* (whose trunk may reach a diameter of 10 m and the tree an age of 2000 years). This species, having great ecological plasticity, can live in other African biomes, and even among mangroves. The Sinai is dominated by the well-known tamarisk *Tamarix mannifera* which provided the biblical manna of the Hebrews. There are three principal kinds of desert plant roots:

a) *Wide, but shallow fascicled systems*, suitable for absorbing the maximum quantity of water during capricious desert storms (e.g. in the giant saguaro cactus *Cereus giganteus*, the largest of all North American cacti).

Fig. 8. *Welwitschia mirabilis*, the famous dwarf gymnosperm endemic to the Namib Desert

b) *Bulbs and other underground reserve structures* (e.g. in the above-mentioned bi of the Bushmen).

c) *Deep primary roots*, present in plants found in alluvial basins (e.g. in the mesquite *Prosopis glandulosa*, whose roots can reach a depth of 30 m to tap subsurface water).

In addition to the phenomena already mentioned of ephemerality, succulence, dormancy, reduction transpiring surface areas by partly or wholly shedding aboveground parts, and appropriate root behaviours, it is worth remembering other important plant adaptations to the xeric conditions of the desert: stomatal closure, reduction in the speed of metabolism and body size, production of protective resins or waxy coverings, maximization of seed dispersal, etc. A particular phenomenon is spinescence: many leaves of desert plants undergo a transformation to thorns. Contrary opinions have been expressed on the function of this adaptation. At first, it was interpreted as a method of reducing transpiration, later as a protection against grazing animals and, more recently (Pond 1962; Hadley 1972; Cloudsley-Thompson 1975), as a means of reducing incident solar radiation and, consequently, the heat transmitted to the surface of plants. A multipurpose function of this adaptation cannot be excluded, as in most other biological phenomena.

All these climatic plant adaptations affect the biology and behaviour of desert animals. Ephemerality imposes a perfect life-cycle synchronization between plants and their exploiters, to achieve the double aim of pollination

Fig. 9. The spinescent cucurbit 'nara' (*Acanthosycios horridus*). *Above* Shrubs; *below* ripe melons, food for various animals and even people, such as the Hottentots

effectiveness and the timely exploitation of ephemeral trophic resources (e.g. Common 1970; Wainwright 1978a, 1978b). Succulence is providential for many animals (Man included) which have difficulty finding drinkable water: it is sufficient to cite the case of the American kangaroo rat *Dipodomys* sp., which never drinks during its lifetime and only occasionally includes juicy grasses and pieces of cacti in a diet mainly based on dried seeds (Ghobrial and Nour 1975). Plant spinescence is very useful to many animals, enabling them to avoid their predators. Several invertebrate species are adapted to live in the protected islands between the spines: e.g., this is the case of the Namib orthopteran *Acanthoproctus diademata*, which inhabits the impenetrable thorny shrubs of the nara, *Acanthosycios horridus* (Fig. 9). Various small vertebrates, mainly birds, are safe among the thorns or within the tissues of succulent plants. The gila woodpecker, *Centurus uropygialis*, digs its nest into saguaro trunks – these nests may later be utilized by elf owls, *Micrathene whitneyi*, as day shelters; the cuckoo *Geococcyx californianus* nests on piles of thorny twigs; the American rat *Neotoma cinerea* protects the entrance to its hole with masses of thorns of cholla, one of the most prickly North American cacti.

3 Desert Zoocoenosis

3.1 General Characteristics

Seemingly uninhabitated deserts provide shelter for surprisingly rich and diverse zoocoenoses. Only in some extreme cases, such as the polar areas, is the faunal component very scanty: in Antarctica, where vegetation essentially consists of blue-green algae, lichens (more than 400 species) and mosses, invertebrates include only protozoans, turbellarian worms, nematodes, rotifers, tardigrades, mites, ostracods, copepods, collembolans and dipterans. On the Antarctic coasts animal life is more abundant, including also various vertebrates (i.e. petrels, penguins, pinnipeds), since nourishment is supplied there by marine resources, which is generally the case on the edges of the sea throughout the whole world.

Wherever producers are present, consumers are found. Of course, this ecological rule is also in force in desert ecosystems, where herbivores, carnivores and detritivores are all well represented, although not easily visible. Only the experienced eye of a specialist can discern the numerous but inconspicuous species, since they are reluctant to carry on external activity under adverse climatic conditions. Most animals spend the daylight hours under rocks or in crevices of the substratum or in dens burrowed by themselves (burrowers are especially prevalent in sandy areas): they are surface-active at night, after sunset or early in the morning. Several camouflage techniques make the discovery of animals even more difficult.

An intricate series of interactions takes place among producers and consumers. They are not limited to trophic aspects. Trees and shrubs offer protection to animals against predators and climatic extremes. In many cases, producers and primary consumers establish stable communities, and several scientists (e.g. Norton and Smith 1975; Mispagel 1978; Crawford 1981) are convinced of their co-evolution: this opinion is supported by the consideration of the perfect synchronization between ephemeral plants and their pollinators. The same rain that stimulates the flowering of annual plants also induces insects to emerge from their cocoons, pupae and other preimaginal stages. This contemporaneity is a prerequisite for the subsequent animal-plant interaction. The latter is further guaranteed by irresistible plant signals, such as showy flower colours or odours which quickly and specifically attract receptive insects, with advantages for both parties: primary consumers have nectar stores at their disposal, and producers will be pollinated. Considering that the activity of insectivores is also synchronized with that of their prey, the close interdependence of all the biocoenotic components becomes obvious.

A particular category of herbivores is represented by seed-eaters. Plants produce large quantities of seeds some of which survive the collection and digestion by rodents, ants, birds, etc., whose populations greatly fluctuate from year to year according to seed abundance. In various cases another interesting interaction arises between plants and herbivores (Leopold 1962). Mesquite pods, for example, contain seeds too hard to be digested; yet the gastric juices of herbivores erode the seed integument, making the penetration of water (which would otherwise be unlikely) possible and so starting rapid germination; in addition, dung represents an excellent medium for the initial growth of embryo plants. In a similar manner, the hard seeds of the Kalahari baobab germinate more quickly after passing through the digestive tract of baboons.

Some animal species may live in desert territories where producers are absent: their survival depends on dried vegetable matter transported by the wind. This food can be so abundant as to be partially buried and exploited even years later (Brinck 1956; Cloudsley-Thompson 1975, 1977a).

Many desert species are specialized for feeding on particular non-animal organisms or parts of them. For example, various nematodes, mites and collembolans are bacteriophages or fungivores; rotifers and tardigrades exploit the cellular contents of mosses and lichens; certain Gastropoda, Orthoptera, Coleoptera, Lepidoptera and Diptera are defoliators; some beetles and thrips are flower feeders; a few Coleoptera, Hemiptera and Lepidoptera are fruit eaters; many ants are granivores; beetles, butterflies, moths, bees, wasps and flies are pollen and nectar feeders; certain termites are wood eaters, and so on.

Several desert animals, belonging to different systematic groups, are detritivores (Sect. 7.3). They play a key role in the processes of litter breakdown and decomposition (Crawford 1979). Some desert arthropods, such as the North American millipede *Orthoporus ornatus* (Wooten and Crawford 1975) and the social isopod *Hemilepistus reaumuri* (Shachak et al. 1976) seem to ingest soil as a standard part of their diet, acting as dispersal agents for fungi and other micro-decomposers. Some insect species are highly specialized for utilizing wind-blown debris as food: the Saharan isopteran *Psammotermes hybostoma* (Harris 1970) and the Namibian tenebrionid *Lepidochora argentogrisea* (Louw 1972) represent excellent examples of this nutritional niche.

Secondary consumers are very numerous: carnivorous species are present in several groups of invertebrates (from nematodes to arachnids, chilopods and insects) and in all the terrestrial groups of vertebrates. In this context, a particular category of predators is represented by insectivorous animals: the great quantity of desert insects provides secondary consumers with conspicuous biomass. In spite of this, cannibalism as been noted in several animal species, even though this phenomenon is more frequent under stressful conditions. Nevertheless, it has been shown (Polis 1979, 1980; Polis and Farley 1979a, b) that homophagy is very frequent in the sand scorpion *Paruroctonus mesaensis* and represents a regular technique of demographic regulation. It is possible that similar control of population size is exercised by other predatory species.

Many other species exploit food matter of animal provenance. There are dung feeders (such as Scarabaeidae, Trogidae and Geotrupidae) (Fig. 10), carrion feeders (including various mites and beetles), and blood suckers (such as various dipterans). Furthermore, various parasitic forms, belonging to different invertebrate taxa (nematodes, mites, etc.), are widespread in all dry environments.

Finally, we must not forget the great diffusion in arid and semi-arid habitats of omnivorous species: the diet of these animals is extremely varied and can include both vegetable and animal food. Good examples of this are furnished by both invertebrates (nematodes, termites, beetles, etc.) and various vertebrates.

Fig. 10. Male of the Fuegian geotrupid beetle *Taurocerastes patagonicus*. The two frontal horns are the most distinctive male feature

3.2 Main Animal Groups

3.2.1 Protozoans

These ubiquitous unicellular animals are found in all deserts (Crawford 1981); however, very little is known of their biology and behaviour in arid and semi-arid regions. They can inhabit both hot and cold deserts, as well as the Antarctic. When humus is present, these organisms are abundant in a layer at a depth of 20–30 cm, as reported by Fuller (1974) with respect to the Central Asian desert. They can also live in sandy areas, provided the surface temperature never reaches 70 °C (at this level, even encysted protozoan forms die). In the Sahara, near Cairo, El-Kifl and Ghabbour (1984) identified various species of ciliates in pure sand, which survive in water membranes produced by the dew deposited daily around sand granules. In the Antarctic, testate amoebae and ciliates live in moss peat; ciliates and flagellates in moraine soils (Wallwork 1976). In temporary waters of the desert, some protozoans also survive; for instance, Rzòska (1961) found ciliates in rain pools in the Sudan.

3.2.2 Turbellarian Flatworms

These are predominantly marine animals; they are often benthic, inhabiting muddy or sandy substrata. Various turbellarians, such as the well-known planaria, live in fresh water; a few species, such as the North American *Bipalium*, are terrestrial and confined to moist environments. Sublette and Sublette (1967) discovered a turbellarian flatworm in temporary desert playa lakes in the Llano Estacado along the New Mexico-Texas border.

3.2.3 Rotifers

This primitive group of pseudocelomates has, as yet, scarcely been investigated. It is known that some fresh water Monogononta can be found in dense masses in temporary desert waters (Rzòska 1961) and that several bdelloid rotifers are adapted to live on mosses and lichens in various regions of the world, Antarctic included. Their incredible ability to endure extremely low temperatures (−272 °C) without damage, and to survive for several years under complete drought (Barnes 1968), deserves the detailed attention of desert biologists.

3.2.4 *Nematodes*

Nematodes are widespread at all latitudes, from polar to tropical regions, and in various habitats. They are also abundant in arid environments. There, particularly concentrated around the roots of plants (Wallwork 1982), they represent a fundamental component of the soil fauna. On the basis of their feeding habits, they have been classified by various authors as fungivores, bacteriophages, detritivores, predators, omnivores and parasites, in order to specify their role in ecological processes. During periods of drought, many soil as well as moss nematodes, which are hydrobiont, can stop their metabolic activity by entering a quiescent state (anhydrobiosis), folding their bodies longitudinally and transversely to minimize water loss (Freckman 1978). Thanks to these characteristics they are able to live in any kind of environment with only a small or temporary supply of water. Nematodes can at times be found in temporary desert waters: Rzòska (1961, 1984) noted the presence of these worms in tropical rain pools.

As far as numbers of individuals are concerned, nematodes are less abundant in arid environments than in moist soils. In coastal dunes, 0.3–2 million nematodes per square meter have been calculated compared to 12–30 million which have been recorded in some forest soils. However, very restricted habitats, such as sand dunes, host a nematode fauna rich in species, most of which are rare or even unknown elsewhere (Yeates 1967; Vinciguerra and Zullini 1980; Marinari et al. 1981).

3.2.5 *Annelids*

Terrestrial and fresh water annelids were once thought to be absent from deserts. This opinion has recently been corrected. Sublette and Sublette (1967) collected oligochaetes and leeches in North American temporary desert waters. Furthermore, considerable data on enchytraeid oligochaetes living in arid environments are progressively being accumulated. Chernov et al. (1977) found a large quantity of enchytraeids in the soil fauna of the Arctic desert; Santos and Whitford (1979) and Springett (1979) noted the presence of these minute, whitish worms in arid regions of North America and Australia respectively. Many other oligochaetes inhabit Saharan territories (El-Kifl and Ghabbour 1984), often transported there passively from neighbouring regions by natural agents (such as logs or floating plants) or by human activities; nevertheless, some endemism is hypothesized.

3.2.6 Gastropods

Pulmonate gastropods are widespread in all terrestrial environments. In spite of their need of moisture, many snail species are also found in deserts: morphological, physiological and behavioural adaptations allow them to survive under arid conditions. Desert species, belonging to various families, have been described from Australia (McMichael and Iredale 1959), North Africa (Cloudsley-Thompson and Chadwick 1964), South America (Jaeckel 1969), the Middle East (Yom-Tov 1970), North America (Bequaert and Miller 1973) and South Africa (Van Bruggen 1978).

3.2.7 Tardigrades

These small invertebrates (their body does not usually exceed 0.5 mm in length) are aquatic. The so-called terrestrial forms live in the water layers held by the stems of mosses and lichens. During periods of drought, like rotifers, they may survive for a long time by entering an anabiotic state (cryptobiosis). They are also able to endure very low temperatures (−272 °C) and prolonged immersion in vinegar, ether, absolute alcohol and other toxic liquids (Barnes 1968).

Because of their extraordinary plasticity, these animals are ubiquitous; several species can inhabit both hot and cold (and polar) deserts. We have very little information on desert tardigrades. The presence of Echiniscidae, Macrobiotidae and Calohypsibiidae in the Sahara and in the Namib has recently been ascertained (Binda and Pilato 1987; Pilato et al. 1991). Some desert tardigrades are cosmopolitan, but the Saharan *Hexapodibius bindae* appears to be endemic (Pilato 1982).

3.2.8 Arachnids

These chelicerate arthropods are widely distributed in the deserts of the world. However, some orders are absent from certain arid environments: only mites occur in the polar deserts, uropygids are absent from the arid areas of Africa and Australia, while solifugids are not found in the Australian deserts. Notwithstanding these exceptions, the arachnids, predominantly predatory animals, represent a peculiar component of arid ecosystems.

Scorpions (probably among the oldest terrestrial arthropods) are symbolic animals of the desert, even though various species inhabit humid environments (Cloudsley-Thompson 1975). Generally, they have nocturnal habits and capture prey with their pedipalps; when necessary, they can kill by stinging at which time they inject a potent poison. Some species produce a neurotoxic substance, lethal also to Man. The Saharan *Androctonus australis*, the Palaearctic *Leiurus quinquestriatus* and the Mexican *Centruroides sculpturatus* belong to this category of disagreeable invertebrates.

Solifugids, fast runners over the ground surface, are very voracious predators. Many species are nocturnal, but a few North American and South African species are day-active. Unlike those of scorpions, solifugid pedipalps are without nippers, but the tip is endowed with an adhesive organ, specialized for catching prey. The huge chelicers kill the victim, tearing its tissues apart.

Uropygids, called whip scorpions because of their flagellarlike terminal opisthosomal segment, are nocturnal predators and, like solifugids, are fast runners. The chelicerae are small, while the pedipalps are well developed and stocky. Uropygids protect themselves from predators by emitting repugnant substances. The North American vinegaroon *Mastigoproctus giganteus* produces a mixture of acetic and caprilic acid (Jaeger 1957), whereas the South Asian *Thelyphonus caudatus* uses formic acid to discourage its attackers.

Spiders are present in all but polar deserts. The orthognath spider families Ctenizidae and Theraphosidae have been recorded in various North American deserts (Gertsh 1949). Several dune-adapted species of Labidognatha, belonging to different families (Sparassidae, Thomisidae, Salticidae, Agelenidae, Lycosidae, Theridiidae, Araneidae, Gnaphosidae, Hersiliidae, Tetragnathidae) are common in many deserts (Cloudsley-Thompson and Chadwick 1964; Newlands 1978; Cloudsley-Thompson 1984). Nevertheless, I do not think that desert spiders have been adequately investigated as yet. For example, I myself, with some collaborators, have recently discovered some new species of Segestriidae (*Ariadna* spp.), inhabiting the gravel plains of the Namib desert. One of them (Fig. 11) is characterized by a ring of seven stones arranged around the holes of their tubular burrows in the sandy substratum (Costa et al. 1993, 1995). Spiders have predaceous habits, feeding on insects and other arthropods. Occasionally, they attack small vertebrates. Their bite is generally harmless to Man, but the poison may sometimes be dangerous to children or to people in poor health. For example, the bite of the notorious black widow (*Latrodectus mactans*) may be fatal to individuals suffering from hypertension or coronary heart disease.

Mites are a conspicuous component of the desert soil fauna. They are common in the hot deserts (Wallwork 1982), but some species may also live in the Antarctic (Wallwork 1976) and Arctic (Chernov et al. 1977) regions. Acarine arthropods may be carnivores, saprophages, detritivores, bacteriophages, fungivores or parasites. Some giant velvet mites (Trombidiidae) show an amazing increase in the adult population a few days after rain: this phenomenon has been described by Cloudsley-Thompson (1962) and Tevis and Newell (1962) in the African and North American deserts respectively. Many ixodid ticks are known to be vectors of disease to Man and domesticated animals.

Fig. 11. The seven-stone ring of the Namibian segestriid spider, *Ariadna* sp. *Below the finger* a hole is under construction

Other arachnid orders are represented in certain deserts. Pseudoscorpions (false scorpions) and Opiliones (harvestmen) are occasionally found in North Africa (Cloudsley-Thompson 1956; Beier 1962; Ghabbour et al. 1977), North America (Hoff 1959; Allred 1965; Weygoldt 1969) and South America (Deboutteville and Rapoport 1968). Very little is known of their biology and their adaptations to desert life.

3.2.9 Crustaceans

Crustaceans are predominantly aquatic mandibulate arthropods. Nevertheless, various groups of this class live in arid and semi-arid environments. Some occur in temporary desert waters; they belong to the phyllopods which include essentially anostracans (fairy shrimps), conchostracans (clam shrimps) and notostracans (tadpole shrimps). Copepods, such as the cyclopoid species *Metacyclops minutus*, which inhabits temporary desert ponds in the Sudan (Rzòska 1961), may also be present. Phyllopods use indefinite periods of dormancy to face unpredictable periods of drought and are capable of instantly exploiting the first rainfall. Reactivation is rapidly initiated and the animals grow to maturity very quickly: this 'explosion' of life, which occurs in a temporary desert pond, represents a striking example of animal ephemerality.

Oniscoid isopods (wood lice) are amphibious or terrestrial crustaceans, widespread in all the moist areas of the world. Nevertheless, some wood lice can live in deserts and include North African, Asian, Australian and North American species (Cloudsley-Thompson and Chadwick 1964; Allred and Mulaik 1965; Warburg 1965a, b; Schneider 1971; Linsenmayr and Linsenmayr 1971; Warburg et al. 1978; Kheirallah 1979). Xeric species of *Hemilepistus* have aroused great interest due to their high level of social organization (Linsenmayr and Linsenmayr 1971; Linsenmayr 1972; Shachak 1980; Sects. 8.4 and 9.1).

3.2.10 Myriapods

Diplopods (millipedes) are detritivorous or herbivorous animals, not well adapted to dry conditions. Yet, some large Spirostreptidae do occur in various arid regions: in the Middle East (Verhoeff 1935; Kaestner 1968), Africa (Kraus 1966; Lawrence 1966) and North America (Loomis 1966; Causey 1975; Crawford 1976). The desert species most investigated is *Orthoporus ornatus*, found in the southwestern United States (Crawford 1972; Riddle et al. 1976; Crawford 1978; Pugach and Crawford 1978). Some polydesmoid diplopods live in semi-arid areas bordering deserts.

Chilopods (centipedes) are carnivorous myriapods, better adapted to desert life. They have been found in North America (Chamberlin 1943; Crabill 1960), North Africa (Cloudsley-Thompson 1956; Cloudsley-Thompson and Chadwick 1964) and South Africa (Lawrence 1975). Desert centipedes belong mainly to the family Scolopendridae; however, some geophilomorphs, but rarely lithobiomorphs (see Cloudsley-Thompson 1958) have also been described from arid environments. Scolopendromorphs, which are the predominating chilopods in hot deserts, can reach a very large size. Moreover, they have nocturnal habits, based upon an endogenous nocturnal activity rhythm which persists in continuous darkness (Cloudsley-Thompson and Crawford 1970). The pauropods, another group of subterranean myriapods, may also occur in some hot deserts. Very little is yet known of their biology (Wallwork 1982).

3.2.11 Insects

This class of terrestrial arthropods is undoubtedly the widest and most varied animal group. The number of known species of insects is at present about

750 000; however, this is probably only a quarter of all existing species. Species number and diversity are proof of the evolutionary success of these animals, which have been able to colonize all terrestrial environments and can also invade aquatic habitats. Several factors have contributed to their success; however, a fundamental advantage is their ability to fly.

The arid and semi-arid ecosystems support a surprising quantity of insect species. Cloudsley-Thompson (1975) estimated that no less than 26 out of a total of 32 orders of this class are present in the NW Sahara. Bird (1930) calculated that the total number of individual insects may reach 3 800 000 ha^{-1} in a steppe environment. Relatively few groups of wingless insects (Collembola and Thysanura among Apterygota, Thysanoptera, Psocoptera and certain Coleoptera among Pterygota) spend their entire life span in subterranean habitats or represent a fundamental ecological component of the soil (Wallwork 1982). Many other insects also pollinate plants. On the whole, they form a remarkable part of the primary consumer fauna; on the other hand, they furnish food for an immense number of insectivorous animals (other invertebrates, reptiles, birds, mammals).

Collembola (springtails) occur in very large numbers in soils throughout the world, from the polar to the hot deserts (Wallwork 1982). Many of these are detritivores, others are fungivores, herbivores or bacteriophages. Thysanura (bristletails) are particularly prone to colonize dry environments (they may even abound in sandy areas); they are active at night and belong to the category of detritivores.

Various species of cockroaches live in arid regions (Roth and Willis 1961). They have been described from the Sahara, the Central Asian, North American and Australian deserts. Little is known of their biology (Wallwork 1982), but the Egyptian *Heterogamia syriaca* (Ghabbour and Mikhail 1977) and the North American *Arenivaga investigata* (Hawke and Farley 1973; Edney and McFarlane 1974; Edney et al. 1974, 1978) have been extensively studied.

The ecology of the lesser orthopteroids living in arid environments has been poorly investigated as yet. We have only limited information on their presence in various deserts: mantids and phasmatids in the Sahara (Chopard 1938; Leouffre 1953) and North America (Tinkham 1948); dermapterans in North America (Helfner 1953), South America and South Africa (Brindle 1984); embiopterans in the Mojave desert (Beck and Allred 1968) and Qatar (Cloudsley-Thompson 1986).

Orthopterans (crickets, locusts and grasshoppers) are the dominant insects of semi-arid areas. There are also several species in non-polar deserts, especially near the vegetation on which they feed; they are particularly abundant after rain. Primary consumers, locusts, often migrate when food is locally insufficient, becoming a real plague of cultivated land (Matthews 1976; Uvarov 1977). Crickets and grasshoppers, dwelling in arid and semi-arid environments, have also been the object of considerable research regarding various aspects of their eco-ethology.

Very common in tropical areas, Isoptera (termites) can be found in the African, North American, Asian and Australian deserts (Lee and Wood 1971; Brian 1978). These eusocial insects (Wilson 1971) construct subterranean nests in the more arid regions (Cloudsley-Thompson 1975); elsewhere, they construct aboveground nests or mounds. These animals may be wood-eaters, harvesters or omnivores: thanks to the activity of symbiotic protozoans, bacteria and fungi, they are able to digest cellulose, chitin and lignin.

Hemipteran, homopteran, thysanopteran and psocopteran insects are frequent in various desert areas, but their ecology is, as yet, little known. Nevertheless, they play important, not always identified, roles in desert ecosystems. Here, only the unique case of the Namibian 'Welwitschia bug', *Odontopodus* (= *Probergrothius*) *sexpunctatus* (Fig. 12), which has been hypothesized as fundamental for the pollination of the Namibian *Welwitschia mirabilis* will be mentioned. Both nymphs and adults of this non-flying pyrrhocorid hemipteran feed exclusively on the cones of females *Welwitschia*. On the other hand, Bornman (1972) maintained wind pollination as more feasible for this dwarf gymnosperm. Marsh (1982) found certain wasps which carry pollen and visit the female cones. According to this author, *P. sexpunctatus* is able to tolerate phenolics produced by the *Welwitschia*

Fig. 12. Adults of the Welwitschia bug (*Odontopodus sexpunctatus*), the most common inhabitant of *Welwitschia mirabilis*

plants as protection against herbivores. The bug may even utilize these chemicals to make it unpalatable to predators, however, there is a reduviid bug which preys regularly upon it (Seely 1987).

Several neuropteran families are represented in non-polar deserts. In sandy arid environments, myrmeleontids (ant lions) are the most common neuropterans; their larvae ('demons of the dust') are well known for their ambush technique of predation: they stay buried at the bottom of cone-shaped pitfall traps waiting for their unfortunate prey (Wheeler 1930).

Coleoptera are the insects best adapted to desert life (Cloudsley-Thompson and Chadwick 1964). This group, the world's largest animal order (over 300 000 species), is widespread in arid and semi-arid environments and is represented by many families. Tenebrionids are the most typical desert beetles, but Scarabaeidae, Geotrupidae, Trogidae, Curculionidae, Cerambycidae, Chrysomelidae, Meloidae, Cicindelidae, Carabidae, etc. are also sometimes conspicuous members of desert faunas. Some are vegetarian, some predatory, others detritivorous or coprophagous; many are omnivorous, feeding on vegetable and animal matter, carrion and dung included. Beetles may have daytime, crepuscular or nocturnal habits. Many of them have a pale body colouration, many others (mainly tenebrionids) are dark-coloured (Fig. 13). Various hypotheses have been formulated to explain the adaptive value of the body colouration: there is fair agreement on the cryptic value of light coloura-

Fig. 13. *Physadesmia globosa*, typical black tenebrionid beetle living in the central Namib. The female is closely followed by a male, as often occurs

tion (Cloudsley-Thompson 1979), while a lively controversy is attached to the adaptiveness of the black colouration. According to some authors, the latter may have an aposematic ('warning') function, since many black beetles are able to liberate noxious chemicals if disturbed (Eisner and Meinwald 1966; Doyen and Somerby 1974; Tschinkel 1975). According to Hamilton (1973), the black colour is instead associated with very high thermal requirements (the 'maxithermy' hypothesis). However, the validity of this hypothesis has recently been questioned (Ward 1991). Alternative explanations for the black colouration of desert beetles are associated with the properties of melanin (waterproofing and/or strengthening of the cuticle; protection against ultra-violet radiation, etc.).

Numerous species of Lepidoptera have been found in the vegetated parts of deserts; but often they are cosmopolitan or migrant, and only relatively few are really endemic. Very little is known about the adaptation of butterflies and moths to desert life.

Diptera is the insect order most widespread in all the terrestrial environments; in fact, we can find flies everywhere, even in the Arctic (Chernov et al. 1977) and Antarctic (Holdgate 1977). Desert flies belong to various families (Muscidae, Psychodidae, Tipulidae, Syrphidae, Nemestrinidae, Mydaidae, Bombyliidae, Asilidae, Therevidae, Empidae, Tabanidae, Oestridae, Rhagionidae, etc.). Evidently having achieved great evolutionary success by improving their flight performance, they have been able to occupy several feeding niches, ranging from plant galls to carrion utilization, from nectar-sucking to blood-sucking, and from dung feeding to predation.

Hymenopterans are widely dispersed in all non-polar deserts. Only a few species of Symphyta dwell in arid environments (Riek et al. 1970), whereas Apocrita are very numerous in the desert. Bees are strictly dependent upon blooming plants (Orians and Solbrig 1977). Consequently, the adults are found only during the rainy season. In contrast, wasps and ants are present at all times and everywhere. Many wasps (Sphecidae, Pompilidae, Scoliidae, Mutillidae, Chrysididae, and Vespidae) paralyze and carry spiders and insects to their burrows as provisions for subsequent generations. Ichneumonoid and chalcidoid wasps lay their eggs directly in the body of the prey. Ants, which are all eusocial (Wilson 1971), usually build collective nests underground, but some species utilize dead trees or bark (Cloudsley-Thompson 1975). Many ants are harvesters, others are predaceous, while still other species have different feeding habits. Undoubtedly, the Formicidae play a fundamental role in desert ecosystems.

3.2.12 Fishes

Since dry regions can include aquatic habitats, fishes may also be considered desert inhabitants. Some of them have been dispersed by Man in arid environments, others are indigenous (Deacon and Minckley 1974) even when major streams originate from higher, non-desert elevations. Fishes living in torrential streams are endowed with adhesive organs, such as sucking discs, hooks, and other structural modifications, to face the impetuosity of unidirectional, turbulent currents (Hora 1922). In isolated water basins, intermittent streams, springs and subterranean water-bearing strata, desert fishes exhibit special adaptations. Since the last glacial era, generations of the Californian Death Valley 'pupfish' (*Cyprinodon milleri*) have become able to endure thermal extremes, from freezing to over 40 °C (Leopold 1962). The Saharan catfish *Clarias lazera* can jump from one muddy pool to another and, when the last drop of liquid has vanished, it can roll into a ball and cease its metabolic activity until the next period of humidity.

3.2.13 Amphibians

Although they are strictly dependent on water for breeding, anurans are widely distributed in many deserts; but only about 3% of approximately 3500 species are able to inhabit very arid regions (Sullivan 1989). Some species enjoy a wide variety of ecological habitats, and may live from sea level (sometimes also below it) to very high altitudes. For example, the Pacific tree frog *Hyla regilla* can be found near Salton Sea (California) as well as in alpine meadows at 4250 m elevation (Jameson 1966). The green toad *Bufo viridis*, which is found in various North African oases, has also been found in the Himalayas at an elevation of 4570 m (Cochran 1961).

Some frogs and toads have colonized permanent waters in different deserts of the world, others may inhabit temporary rain pools. They avail themselves of several morphophysiological and behavioural adaptations to avoid desiccation. Some species of the genus *Scaphiopus* (spadefoot toads) have been studied especially, since they live in more arid regions (Mayhew 1968; Blair 1976).

Certain urodeles, such the tiger salamander (*Ambystoma tigrinum*), can also be found in desert environments (Delson and Whitford 1973).

3.2.14 Reptiles

In contrast to amphibians, xeric habitats are congenial to reptiles. These animals appear actually to be predisposed to face life in arid environments, as is demonstrated by the great abundance of reptilian species in the desert, belonging to several taxonomic groups. Tortoises, lizards and snakes are able to minimize water loss through their body surface, which is covered by scales and lacks exocrine glands. In addition, they benefit from uricotelic excretion. Several thermo-hygroregulative behavioural adaptations provide further evidence of the great evolutionary success of reptiles in all but polar deserts. At least three species of Squamata may reach areas near the Arctic Circle: the viviparous lizard *Lacerta vivipara*, the common adder *Vipera berus* and the garter snake *Thamnophis sirtalis*. In the Southern Hemisphere, the range of the iguanid *Liolaemus magellanicus* extends to the extreme limits of South America, including Tierra del Fuego. Some poisonous Elapidae, such as the tiger snake (*Notechis* spp.), are found in Tasmania. As to altitude, the absolute record belongs to a rattlesnake, *Agkistrodon himalayanus*, which lives in the western Himalayas at an elevation of 4880 m; the lizard *Liolaemus multiformis* can be found in the southern Andes of Peru above 4500 m (Pearson 1954; Bellairs 1969).

Only a few species of tortoises have been described from arid environments. The most studied are the North American desert turtles. In sunny areas, these diurnal animals walk fast using their long legs, reaching a speed of 0.36 km/h. According to Leopold (1962), the desert tortoise *Gopherus agassizi* is able to transform part of its vegetable food into water which is stored in two intestinal sacs located under carapace. This water supply (about 0.5 l) allows the turtle to survive throughout the entire dry season. More recently, different explanations have been given (Louw and Seely 1982). Other thermohygric adaptations of desert tortoises will be described later.

Lizards (Sauria) are undoubtedly the most abundant vertebrates in arid habitats. Many systematic groups included in this heterogeneous and perhaps polyphyletic taxon are well represented on all continents. Cloudsley-Thompson (1977a) pointed out that agamids and iguanids have respectively colonized the deserts of the Old and the New World, occupying the same ecological niche in the two different geographical areas. From a biogeographical point of view, very interesting is the presence of certain iguanids, such as the sand iguanid *Chalarodon madagascarensis*, as well as the presence of many agamids, such the moloch, *Moloch horridus*, and various species of the genus *Amphibolurus*, in Australia, but the absence of agamids from Madagascar. From an ecological point of view, no less interesting is the parallel evolution which has allowed desert species, not even closely related taxonomically, to acquire similar morphological and behavioural adaptations to similar environments (e.g. Pianka 1985, 1986; Cloudsely-Thompson 1991). For example, an ecological equivalent of the Australian ant-eating moloch (Pianka and Pianka 1970) is the North American horned iguanid lizard *Phrynosoma platyrhinos* (Pianka and Parker 1975); in an analogous way, the North American lizard-eating leopard iguanid *Gambelia* (= *Crotaphytus*) *wislizenii* (Montanucci 1965, 1967) is behaviourally convergent with the Australian varanid *Varanus eremius* (Pianka 1982), and so on.

The Chamaeleonidae are present in Africa, Madagascar, southern Spain, India and Ceylon. Only a few species, such as the common chameleon *Chamaeleo chamaeleon chamaeleon* (Doumergue 1900), the flap-necked chameleon *C. dilepis* (Fitzsimons 1935), the Namibian *C. namaquensis* (Burrage 1973) and the basilisk chameleon *C. africanus* (Lambert 1984) dwell in arid territories. The Gekkonidae are very common in the Palaearctic and Afrotropical regions, but they can also be found in North America (for example, the banded gecko *Coleonyx variegatus* inhabits southwestern Utah, Hardy 1944), India (for example, the Indian house lizard *Hemidactylus flaviviridis*, Rao 1924), and Australia (for example, various species of the genera *Gehyra* and *Diplodactylus*, Bustard 1964a, b). Some Xantusidae live in the arid regions of North America. The Lacertidae are widespread in the Asian and African deserts, as well as in Mediterranean sandy/coastal areas. The Scincidae frequent various arid and semi-arid environments of Africa, Asia, Europe and America. Some species of Teiidae, belonging to the genus *Cnemidophorus* (whiptail lizards), dwell in dry North American habitats. Some Helodermatidae, for example the Gila monster *Heloderma suspectum* and the beaded lizard *Heloderma horridum* characterize the American Sonoran and Chihuahuan deserts respectively. In addition, some Varanidae can be found in arid and semi-arid areas of Africa, Asia and Australia.

Snakes are also well represented in the world's deserts. Various species with a wide ecological tolerance may live in both humid and dry habitats; but other ophidians, such the North American colubrid *Chionactis occipitalis* (Miller and Stebbins 1964), are strictly desert dwellers. Some Leptotyphlopidae and Typhlopidae (worm snakes) have been found in Asian (*Typhlops braminus*, Minton 1966) or southern African (*Typhlops schlegelii*, Fitzsimons 1962) deserts. *Leptotyphlops humilis* (Klauber 1940; Stebbins 1954) has been described from western North America. A few small species of Boidae, belonging to the genus *Eryx* (sand boas), are spread throughout the dry regions of North Africa, Asia and southeastern Europe. Many species of Colubridae, belonging to different genera, inhabit arid areas in America, Asia and Africa. Similarly, the Elapidae (cobras and coral snakes) and the Viperidae (both Viperinae or true vipers and Crotalinae or pit vipers) are present in various desert regions. Elapids are well represented in dry territories of Australia; the Sonoran coral snake *Micruroides euryxanthus* is the only venomous elapid found in American deserts (Minton 1968). The Australian death adder *Acanthopis antarcticus* (Elapidae) shows convergence with the Viperidae. Vipers (for example, the Afro-Asian saw-scaled viper *Echis carinatus*, the North African desert horned viper *Cerastes cerastes*, the South African *Bitis peringueyi*, etc.) are confined to the Palaearctic, Afrotropical and Australian regions, but they (as well as Elapidae) are absent in Madagascar. Pit vipers are widespread in the New World (a representative example is the Mojave rattlesnake *Crotalus scutulatus*), but they can also be found in the eastern and central Asia.

3.2.15 Birds

This class of vertebrate is well represented in the arid regions of the world. Among the Palaeognata, cursorial but flightless birds, ostriches (*Struthio*

Fig. 14. Ostrich (*Struthio camelus*) on the look out

camelus) (Fig. 14) are certainly the more representative inhabitants of deserts and semi-deserts due to both their large size (they can reach 2.5 m in height) and their behaviour; they are widespread in the arid and semi-arid regions of Africa, Arabia and Syria. The smaller South American rhea (also called nandu) *Rhea americana* and the Australian emu *Dromaius novae-hollandiae* present almost similar morphological and ethological features, but they are confined to steppe environments. In addition, some brachypterous Tinamiformes can be found in the Argentine pampas.

The Neognata, comprising the majority of living birds, are abundantly represented in deserts. The flightless but able swimmers, Sphenisciformes, are common inhabitants of Antarctica, New Zealand, Tasmania and the most southern parts of Australia, America and Africa. The cosmopolitan order Falconiformes has various arid-adapted species, belonging to the genus *Falco*. The American kestrel *Falco sparverius* is a common resident of the western North American deserts (Bartholomew and Cade 1957a). The peregrine falcon *Falco peregrinus* and the sooty falcon *Falco concolor* can be found in arid environments of the Middle East and North America respectively (Bartholomew and Cade 1963).

Few Galliformes are arid-adapted; among those that are, there is the North American Gambel's quail *Lophortyx gambelii*, which can be found from Nevada to Mexico (Vorhies 1928; Gorsuch 1934; Lowe 1955; Guillion 1960). The Gruiformes and Charadriformes include some desert species, such as the resident Saharan houbara bustard *Chlamydotis undulata* and the cream-coloured courser *Cursorius cursor*. Various Pteroclidiformes (sand grouse), belonging to the genera *Pterocles* and *Syrrhaptes*, are present in the African, southern Mediterranean, eastern Asian and Indian arid areas. In particular, African sand grouse have been studied extensively (Heim de Balsac 1936; George 1969, 1970; Dixon and Louw 1978). Some Columbiformes may also inhabit deserts in various regions of the world. The white-winged dove *Zenaida asiatica*, the inca dove *Scardafella inca*, the mourning dove *Zenaidura macroura marginella* and the ground dove *Columbina passerina*, which live in the North American deserts, have been the object of detailed research on body temperature and water requirements (Bartholomew and Dawson 1954; Bartholomew and MacMillen 1960; MacMillen 1962; MacMillen and Trost 1966; Willoughby 1966). Other doves, such the turtle dove *Streptotelia turtur* and the palm dove *Streptotelia senegalensis*, have been recorded to breed regularly in the Sahara (Casselton 1984).

A few Piciformes occur in the arid zones. One species, mentioned above, is the North American gila woodpecker (Sect. 2.5). Among the Cuculiformes, there is the interesting case of the North American cuckoo or roadrunner *Geococcyx californianus*. Calder (1968) studied its diurnal activity, while Calder and Schmidt-Nielsen (1967) analyzed its thermoregulation. An arid land species, belonging to the Psittaciformes, is the Australian budgerigar *Melopsittacus undulatus* (Cade and Dybas 1962). Well-known Strigiformes of arid North American environments are the screech owl *Strix flammea* and the elf owl *Micrathene whitneyi* (Sutton and Sutton 1966; Ligon 1968). The small owl *Athene noctua* was noted by Casselton (1984) as being a Saharan breeding species. The dwarf owl *Glaucidium perlatum* is common in African savannahs. Some Caprimulgiformes, including the Egyptian nightjar *Caprimulgus aegyptius* and the Nubian nightjar *Caprimulgus nubicus*, maintain a more or less stable relation with the North African desert areas (Cloudsley-Thompson 1968; Casselton 1984).

The Passeriformes is the richest order of all, including over half of all existing species. Passeriformes are very numerous in deserts (Fig. 15), but only some of the most representative will be mentioned: finches (trumpeter finch *Rhodopechys githaginea*; scaly-feathered finch *Sporopipes squamifrons*; zebra finch *Taeniopygia castanotis*), warblers (desert warbler *Sylvia nana*; streaked scrub warbler *Scotocerca inquieta*), larks (sand lark *Ammomanes deserti;* Gray's lark *Ammomanes grayi*; crested lark *Galerida cristata*; bifasciated lark *Alaemon alaudipes*), martins (house martin *Delichon urbica*; pale crag martin *Hirundo obsoleta*), swallows (blue and white swallow *Pygochelidon cyanoleuca*), buntings (Albert's towhee *Pipilo alberti*; house bunting *Emberiza striolata*), shrikes (great grey shrike *Lanius excubitor*), thrushes (blackstart *Cercomela melanura*; fulvous babbler *Turdoides fulva*; Arabian babbler *Turdoides squamiceps*), wheatears (*Oenanthe deserti* and other species of the same genus), sparrows (desert sparrow *Passer simplex;* black-throated

Fig. 15. Desert bird searching for insects

sparrow *Amphispiza bilineata*), ravens (fan-tailed raven *Corvus rhipidurus*; brown-necked raven *Corvus ruficollis*), and many others.

3.2.16 Mammals

Even though less preadapted than birds to the desert, mammals, both herbivorous and carnivorous, inhabit arid and semi-arid territories. Among the Marsupialia, various species widespread in Australia may be considered arid-dwellers. The didelphid Patagonian opossum *Lestodelphis halli* is found further south than other marsupials: it has been found in the desert of southern Patagonia and shows terrestrial and predaceous habits (Thomas 1921). Various Dasyuridae, belonging to different genera, inhabit the central Australian Desert. Of these, the mulgara *Dasycercus cristicauda* and marsupial jerboas of the genus *Antechinomys* are considered to be specialized in desert conditions (Iredale and Troughton 1934; Schmidt-Nielsen and Newsome 1962). The sand-swimming marsupial mole *Notoryctes typhlops* (Notoryctidae) and the long-eared rabbit bandicoot *Thylacomys lagotis* (Peramelidae) are also typical inhabitants of the Australian deserts, including also the red kangaroo

Macropus rufus, the quokka *Setonix brachyurus* and the desert rat-kangaroo *Caloprymnus campestris* (Macropodidae), which occur in arid grasslands (Walker 1968).

Many orders of Placentalia, from Insectivora to Primates, include desert-dwellers. Insectivora have a wide distribution. They are absent only in Australia, New Guinea and the polar regions. Some golden moles, belonging to the African family Chrysochloridae, live in sandy areas; for example, the subspecies *Eremitalpa granti namibensis* inhabits the Namib sand dunes, while *Cryptochloris wintoni zyli* dwells in the coastal sand dunes of western Cape Province, South Africa (Meester 1964). Among the Erinaceidae, the desert hedgehogs (genus *Paraechinus*) and the long-eared desert hedgehogs (genus *Hemiechinus*) inhabit the Afro-Asian deserts and other arid areas; the same is true of the desert shrews (Soricidae) *Notiosorex crawfordi* and *Notiosorex gigas* (from southwestern North America and Central America respectively), *Crocidura smithi* (southern Africa) and *Diplomesodon pulchellum* (Asia) (Walker 1968).

Numerous bats (Chiroptera) are found in arid areas in various regions of the world, even though they always need water daily to satisfy their thirst. The chiropteran species usually have a very wide distribution. The families best represented in desert zones are the Vespertilionidae (vespertilionid bats) of the suborder Microchiroptera and the Pteropodidae (flying foxes) of the suborder Megachiroptera. Some other families, such as Rhinopomatidae, Emballonuridae, Hipposideridae, and Desmodontidae, include arid-dwelling species.

Various Edentata, namely Dasypodidae, inhabit open areas, such as savannahs and pampas; some species, such as the Argentinean pichi *Zaedyus pichyi* and pink fairy armadillo *Chlamidophorus truncatus* can be found only in warm, dry, sandy plains where thorn bushes and cacti are plentiful (Walker 1968). Analogously, the African aardvark *Orycteropus afer*, belonging to the order Tubulidentata, is common in grasslands, where many termites and ants are present. Some non-arboreal Pholidota also live in the semi-arid environments of Africa and Asia, but they are absent from deserts. In contrast, Lagomorpha include several arid-adapted species. The Arctic hare *Lepus arcticus* and the Alaskan hare *Lepus othus* inhabit the Arctic desert; the jackrabbits (*Lepus californicus* and other species of the same genus), the pygmy rabbit *Sylvilagus idahoensis* and the desert cottontail *Sylvilagus auduboni* have colonized the most arid lands of North America.

Numerous species, belonging to the nearly worldwide order Rodentia, are semi-desert or desert-dwellers. They can be found in Arctic regions (collared lemming, *Dicrostonyx torquatus*) and in tundra areas (true lemming, *Lemmus lemmus*), as well as in subantarctic zones (tuco-tucos, *Ctenomys* spp.). They also dwell in steppes, savannahs, pampas, arid grass prairies and mountain plateaux. Some of them live at a very high altitude: the habitat of the Andean chinchilla *Chinchilla laniger* may reach an elevation of 6000 m. Desert species prefer sandy areas or barren rocky places; some of them have colonized gravel plains, others may inhabit clay or saline arid environments. Cricetidae is doubtlessly the family of arid-adapted rodents best represented. Asian cricetid species belong to various genera such *Calomyscus, Phodopus, Ellobius, Meriones, Brachiones, Rhombomys;* African species belong to several genera, among which *Mystromys, Monodia, Gerbillus, Desmodillus, Pachyuromys, Meriones, Sekeetamys,* and *Psammomys* should be cited. On the American continent there are many desert species belonging to the genera *Peromyscus*, *Onychomys, Akodon, Notiomys, Eligmodontia,* and *Neotoma.* Numerous other rodent families include desert species: the Sciuridae (for example, the Asian *Spermophilopsis leptodactylus* and the South African *Xerus inauris*), Geomyidae (the North American pocket gophers *Geomys, Thomomys* and *Cratogeomys*), Heteromyidae (the North American desert pocket mouse *Perognathus penicillatus*, kangaroo mice *Microdipodops*, Merriam's kangaroo rat *Dipodomys merriami*), Muridae (the Australian kangaroo mouse *Notomys filmeri*, the South African gerbil mouse *Malacothryx typica*), Dipodidae (the Afro-Asian desert jerboa *Jaculus jaculus* and various other jerboas belonging to genera *Dipus, Paradipus, Stylodipus, Allactaga, Euchoreutes*), Pedetidae (the South and East African springhaas *Pedetes capensis*), Selevinidae, Erethizontidae, Caviidae, Chinchillidae, Octodontidae, Ctenomyidae, Echimyidae, Petromyidae, Bathyergidae (including the African naked mole rat *Heterocephalus glaber*), and Ctenodactylidae.

Today, Carnivora are found in almost all terrestial environments, being absent only from the Antarctic and some oceanic islands. In Australia and New Zealand, they were introduced by Man (dingos, foxes and other species). They also inhabit the Arctic regions (e.g. the Arctic fox *Alopex lagopus* and the polar bear *Thalarctos maritimus*), as well as areas of southern America where the climate is extreme. Here, the Patagonian weasel *Lyncodon patagonicus* and the Patagonian skunk *Conepatus humboldti* are found. The Falkland Island dog *Dusicyon australis* has been exterminated by Man in the recent past (about 1876), while the Siberian husky (Fig. 16) is even reared in subantarctic areas. Some species of Carnivora may live at very high altitudes. These include the Fuegian fox *Pseudalopex culpaeus lycoides*, the Andean mountain cat *Felis jacobita* and the Central Asian snow leopard *Panthera uncia*, whose ranges may reach elevations of 5000 m. All seven families of Carnivora are represented in desert and/or semi-desert areas. Among the Canidae, it is appropriate to mention foxes (genera *Vulpes, Fennecus, Urocyon, Dusicyon, Alopex, Pseudalopex, Otocyon*), jackals and coyotes (*Canis* spp.), as well as hunting dogs (*Lycaon pictus*). Among the Ursidae, in addition to the polar bear mentioned above, there is the sloth bear *Melursus ursinus*, which can be found in parts of the Indian Desert. One species of Procionidae, the North American cacomistle *Bassariscus astutus*, inhabits broken rocky areas. Various Mustelidae are desert inhabitants: e.g. the marble polecat *Vormela peregusna* (Gobi desert), the North African striped weasel *Poecilictis libyca* (North African deserts), the American badger *Taxidea taxus*, the

Fig. 16. Siberian husky, native of northeastern Asia, reared at Sierra Martial (Tierra del Fuego)

striped skunk *Mephitis mephitis* and the spotted skunks *Spilogale* spp. (North American deserts). Among the Viverridae, besides the genet *Genetta genetta*, which inhabits African and Middle Eastern arid areas, the suricate or meerkat *Suricata suricatta*, the yellow mongoose *Cynictis penicillata* and the dwarf mongooses *Helogale undulata* (grasslands and dry open regions of southern Africa) and *Helogale parvula* (areas of dry acacia brush of northeastern Africa) must be mentioned. The Hyaenidae occur in plains and bushland. For example, the aardwolf *Proteles cristatus* is quite common in open sand plains or bush country of southern and eastern Africa. Moreover, some species of the genus *Hyaena* wander through the African and Asian deserts. Among the lynxes, the Afro-Asian caracal *Lynx caracal* and, among the cats, the Chinese desert cat *Felis bieti* and the American puma *Felis concolor* live in desert environments. In addition, various felids, such as the American jaguarondi *Felis yagouaroundi*, the Old World lion *Panthera leo*, and the Afro-Asian cheetah *Acinonyx jubatus*, inhabit dry open areas.

Both living species of Proboscidea thrive in a variety of habitats and also occur in semi-arid and arid environments where stretches of water are present. The Indian elephant *Elephas maximus* can be found in thick jungle areas to open grassy plains; the African elephant *Loxodonta africana* (Fig. 17) in dense forest to savannah and desert scrub. Some terrestrial Hyracoidea have

Fig. 17. A matriarch African elephant (*Loxodonta africana*) leading her family unit

colonized open grassy lands, rocky territories and arid scrub. In particular, the rock hyraxes *Procavia capensis* and *Heterohyrax syriacus* inhabit rocky and scrub-covered areas of Africa and southwestern Asia respectively; while the Saharan hyrax *Procavia ruficeps* is confined to grassy habitats on the large massifs of the Sahara.

The families Equidae and Rhinocerotidae, belonging to the order Perissodactyla (odd-toed ungulates), are well represented in dry environments. The common zebra *Equus burchelli* is found in Africa south of the Sahara, Grevy's zebra *Equus grevyi* in the open plains of northern East Africa, and the wild horse *Equus przewalskii* in the inhospitable plains of the Altai Mountains in central Asia. The great Indian rhinoceros *Rhinoceros unicornis* can be found in the grassland areas of Nepal, Bengal and Assam; the white rhinoceros *Ceratotherium simum* and the black rhinoceros *Diceros bicornis* dwell in savannah and brush areas of Africa. Still more abundant in desert zones are the Artiodactyla (even-toed ungulates). The American collared peccary *Tayassu tajacus* (Tayassuidae) inhabits very dry lands, while the African wart hog *Phacochoerus aethiopicus* (Suidae) is common in savannah plains. The Camelidae include typical arid-adapted animals. The bactrian camel *Camelus bactrianus* inhabits the cold deserts of central Asia, the Arabian camel or dromedary *Camelus dromedarius* the hot deserts of North Africa and southwest Asia. The guanaco *Lama guanicoe* (Fig. 18), the domesticated llama *Lama peruana* and alpaca *Lama pacos* can be found in semi-desert and high-altitude plains of South America, while the vicuna *Vicugna vicugna* is common in the Andean semi-arid plateaux at elevations between 3500 and 5750 m. Among the Cervidae, the pampas deer *Blastoceros campestris* and the mule deer inhabit dry open plains of South and North America respectively; the caribou *Rangifer tarandus arcticus* and the reindeer *Rangifer tarandus tarandus* inhabit the Arctic regions of the world. The Giraffidae and the Antilocapridae respectively are represented in the arid zones by the giraffe *Giraffa camelopardalis* which is found in the dry savannahs of Africa south of the Sahara, where there are abundant scattered acacia trees, and by the pronghorn *Antilocapra americana* which inhabits the rocky deserts of North America. The Bovidae are widespread in dry environments. Several antelopes, such the greater kudu *Tragelaphus strepsiceros* (Fig. 19), the eland *Taurotragus oryx*, the oryxes (scimitar-horned oryx *Oryx tao*, beisa *O. beisa* and gemsbok *O. gazella*), the addax *Addax nasomaculatus*, the

Fig. 18. The guanaco *Lama guanicoe* is a typical inhabitant of the Chilean part of Tierra del Fuego. In the background, the Strait of Magellan is visible

Fig. 19. Males of the greater kudu *Tragelaphus strepsiceros* are recognizable by their horns

blesbok *Damaliscus dorcas*, the hartebeest *Alcelaphus buselaphus*, the wildebeests [brindled gnu *Connochaetes taurinus* (Fig. 20) and white-tailed gnu *Connochaetes gnou*], the dik-diks (various species of the genus *Madoqua*), the beira *Dorcatragus megalotis*, and the impala *Aepyceros melampus* are widespread in many desert and semi-desert regions of Africa. Other antelopes, including the Arabian oryx *Oryx leucoryx*, the black buck *Antilope cervicapra*, the saiga antelope *Saiga tatarica*, the goral *Naemorhedus goral*, the Tibetan antelope *Pantholops hodgsoni*, etc., inhabit arid and semi-arid areas of Asia. The musk ox *Ovibos muschatus* inhabits the windswept tundras and snowfields of Alaska, Canada and Greenland. In an analogous way, several gazelles (the gerenuk *Litocranius walleri*, the dibatag *Ammodorcas clarkei*, the Persian gazelle *Gazella subgutturosa*, the springbok *Antidorcas marsupialis*, Thomson's gazelle *Gazella thomsoni*, Grant's gazelle *Gazella granti*, etc.), goats and sheep inhabit dry grassy areas.

To conclude this array of more or less arid-adapted animals the order Primates will be discussed. Monkeys are essentially forest-dwellers. Nevertheless, some Old World monkeys (Cercopithecidae), such as baboons and certain macaques, have been able to colonize open grassy plains and rocky areas. For example, the grass monkey *Cercopithecus aethiops* inhabits the African grasslands (it is present with many races, the most important of which are the West African green monkey, the East African grivet

Fig. 20. A wildebeest *Connochaetes taurinus* in Etosha Park (Namibia)

monkey and the South African vervet monkey); the gelada baboon *Theropithecus gelada* the shrubby mountains of Ethiopia; the hamadryas baboon *Comopithecus hamadryas* the rocky hillsides of northeast Africa and Arabia; the anubis baboon *Papio doguera* savannah areas and some of the Saharan massifs; the chacma baboon *Papio ursinus* rocky open country in South Africa and Namibia; and the patas monkey *Erythrocebus patas* the dry grasslands of western and central Africa. A troop of chacma baboons lives in the seasonally dry Kuiseb River canyon (Hamilton et al. 1976; Hamilton 1985). Hamilton (1986) and Brain (1988, 1990) studied the drinking behaviour of these Namib Desert baboons. Among the macaques, the Southeast Asian stump-tailed macaque *Macaca speciosa* lives at high altitudes and is a cold-adapted animal and the Japanese macaque *Macaca fuscata* is confined to the cold island of Honshu. Some macaques, whose habitat can include ocean beaches, are formidable predators of crabs and bivalves.

As a result of his very high adaptability and mobility, greatly increased by an extraordinary cultural evolution, Man (*Homo sapiens*) has succeeded in colonizing the entire world except Antarctica. He can also survive on the Arctic ice and in other cold desert territories. The Eskimos inhabit Greenland, the Aleutian Islands, Alaska, Canada and Siberia. Bushmen (or San) and Hottentots (or Khoikhoi), once widespread in most parts of Africa, are today confined to Botswana, Namibia and the Kalahari; the principal Saharan peoples are the Tubu (or rock people), the Tuareg and the Moors. The Pintupi and the Warlpiri are typical desert Australoids; the Yaghanes, Alacalufes and Onas are virtually extinct human tribes of Tierra del Fuego.

4 Thermohygric Regulation

The main problem to be solved by desert animals is thermohygric regulation. Extreme temperatures and very wide thermal fluctuations, as well as scanty and unforeseeable rainfall, impose special adaptive responses on all the animals inhabiting arid environments. The solutions of this problem may vary considerably and consist in morphological, physiological and behavioural adaptations, which co-operate to allow faunal survival.

Body size and shape can be very important in relation to thermal stresses. Body size is certainly a determining factor for thermal exchange between an organism and its environment. According to the relationship between surface area (S) and body weight (W), $S^3 = W^2$, as the size (and consequently the weight) of the organism increases, its relative surface area decreases proportionally (Louw and Seely 1982). By virtue of this allometric equation, the smallest species can maximize the body surface exposed to solar radiation, but they must protect themselves against nocturnal heat loss. Small mammals have a thick and long pelage, which reduces the speed of heat loss at night and attenuates the daytime heat gain. On the other hand, small animals can easily escape thermal extremes by having access to more favourable microenvironments. In contrast, in the largest species the smaller relative surface area slows down both the daytime heat gain and nocturnal heat loss. The possibility of cooling the body is much reduced and presents a great problem for large herbivores which store considerable amounts of the heat engendered by locomotion. (It should not be forgotten that these animals cannot avoid solar radiation.) For this reason, in large ungulates living in hot climates, there is a clear evolutionary trend toward a shortening and thinning of the hair: the oryx has a very short coat, the rhinoceros and the elephant are almost hairless. The circulatory system also contributes in reducing the negative effects of overheating. The addax utilizes a system of thermal exchange: in the nasal sinuses, the cooled venous blood lowers the temperature of the arterial blood which reaches the brain. Consequently, the animal can cover some hundreds of kilometres in search for food, without its brain overheating. The same system of thermal exchange is adopted by other large antelopes, for example the eland and oryx (Taylor 1969).

In hot, dry environments, various morphological adaptations, evolved with regard to particular organs, play important roles in thermoregulation. For example, various mammals have big ears which are useful for maximizing hearing and also essential for cooling the body. The enormous auricles, endowed with a thick net of tiny and superficial blood vessels, act as radiators. This is especially marked in the North American jackrabbit *Lepus californicus* (Schmidt-Nielsen et al. 1965), but many other desert mammals adopt the same thermoregulative system (e.g. the Saharan fennec *Fennecus zerda*, the North American kit fox *Vulpes macrotis*, the Gobi long-eared jerboa *Euchoreutes naso*, the Australian rabbit bandicoot *Thylacomys lagotis*, etc.).

Several other morphological features assist in the protection from thermal stress. Very long legs occur in both invertebrates and vertebrates. These allow day-active animals to hold the body as far as possible from the hot ground surface during the hottest part of the day. The light colouration of most desert animals, so useful for concealment (homochromy), is also very important in reflecting radiation and so preventing overheating.

Paradoxically, the main physiological system of body cooling is based upon the evaporation of water, which without doubt is the most precious resource in the arid environments. Various techniques of cooling by evaporation are employed by different animal groups. Some large mammals sweat profusely. Small mammals do not sweat but, like reptiles and birds, exploit thermal panting; in particular, birds avail themselves of gular fluttering. Several animals may utilize saliva for evaporative cooling. When the body temperature exceeds 40 °C, tortoises moisten their head and neck with a salivary froth (Cloudsley-Thompson 1970, 1974; McGinnis and Voigt 1971; Riedesel et al. 1971); some Australian marsupials, such as the quokka, lick their legs and moisten them with saliva (Bartholomew 1956). Sometimes, urine is also utilized to lower the body temperature: for example, some tortoises discharge the contents of the cloaca to moisten their hind legs (Cloudsley-Thompson 1970; Riedesel et al. 1971; Sturbaum and Riedesel 1974).

An important series of morphophysiological adaptations concerns the problem of water regulation in desert animals. These, like other organisms, need substantial reservoirs of water although, in arid environments, it is not always possible to find free water to drink. Nevertheless, water is indispensable for cooling. Furthermore, a certain amount of water must be used for the removal of undigested material and for excretion. Several desert animals, both phytophagous and carnivorous, exploit a diet rich in water, and drink irregularly when opportunities occur. The body surface may also function as an almost impermeable barrier to water. This is achieved by the cuticle of arthropods, the scales and skin of reptiles, the feathers of birds and the hairy coat of mammals.

A number of studies on the water economy of various desert animals have provided considerably to our knowledge of their physiological adaptations. Both arthropods and vertebrates are able to reabsorb most of the water from the faecal material during its passage through the hindgut. Camel dung can be utilized for feeding a fire, shortly after its voiding. The faeces of many reptiles are little more than masses of dry dust. The excretion of desert animals can take place with very little water loss. Insects, reptiles and birds excrete uric acid which is insoluble in water and can be easily concentrated. The problem is more difficult for mammals which excrete urea. Because of its great solubility, this inevitably involves a certain amount water loss. Desert mammals with a high rate of evaporation and abundant excretion must drink every day. For instance, bats must go daily to and fro between their shelters and the nearest water source. In contrast, other animals capable of minimizing water loss by evaporation, excretion, defecation and respiration, may spend their entire life without ever drinking (Chew 1965). The survival of these animals depends on the ability to satisfy water requirements by the extensive exploitation of the moisture contained in the food and the metabolic water obtained from oxidation of the food itself (metabolic water production depends on the presence of hydrogen in the food). Examples include insectivorous grasshopper mice (*Onychomys* spp.) (Schmidt-Nielsen and Haines 1964), the predaceous marsupial mulgara (Schmidt-Nielsen and Newsome 1962), kangaroo rats (*Dipodomys* spp.) (Schmidt-Nielsen 1958) and Gray's lark (Ricklefs 1974).

Another important morphophysiological adaptation to desert life, found in various herbivorous mammals, is the ability to store fat in different regions of the body. In the Asiatic camel, fat is stored in the two humps, in Indian zebu cattle *Bos indicus* in the anterior hump. Only one hump, representing the posterior hump of the bactrian camel, is present in the dromedary. Fat is stored here, while in the blackhead Persian sheep fat is deposited on the rump (Fig. 21). (This kind of fat storage resembles the steatopygia of the south-western African Khoikhoi and San women). In North African and Middle Eastern sheep, fat is stored in the large tail. Fat storage, which takes place when juicy food and drinking water are available, provides an invaluable resource for arid-adapted herbivores. Fat contains a large number of hydrogen atoms which, combined with the oxygen atoms originating from respiration, produces energy and metabolic water during prolonged periods of food and water shortage.

Other physiological responses, although not exclusively found in arid environments, may succour desert animals. The anhydrobiosis of nematodes and cryptobiosis of tardigrades, mentioned above, as well as the diapause of temporary pond inhabitants, are efficient systems to surmount climatic extremes. A widespread mechanism for saving energy is torpor, when an animal reduces its metabolic rate to a minimum. Torpor can occur daily according to a circadian rhythm and helps to define the activity rhythm of an animal species. The phenomenon of daily torpor has been extensively studied in small rodents such as the pocket mouse *Perognathus californicus* (Tucker 1965a, b, 1966) and the cactus mouse *Peromyscus eremicus* (MacMillen 1965). Prolongation and deepening of the state of torpor lead to the phenomenon of dormancy, which may last for a season. Summer dormancy or aestivation and winter dormancy or hibernation are well-known physiological states, during which

Fig. 21. The arid-adapted blackhead Persian sheep (*Ovis steatopyga persica*). Native of Persia, this animal is reared in several African areas

oxygen consumption, metabolic rate and body temperature are markedly reduced. Hibernation, common to both invertebrate and vertebrate animals living in cold climates, generally occurs when the mean seasonal temperature falls below 15 °C. Aestivation has been described in several ectothermal and endothermal vertebrates living in hot arid environments and other tropical biomes. A number of ethological adaptations co-operate with morphological features and physiological mechanisms in enabling desert animals to achieve thermohygric regulation.

4.1 Timing of Activity Rhythms

The choice of a temporal niche is fundamental for the ethological characterization of desert animals (Fig. 22). Thermophilous species opt for daytime habits, hygrophilous species are preferentially nocturnal. The nightly drop in temperature radically changes environmental features. Species unable to withstand the night-time cold cease their activity at dusk and take shelter whenever possible. Sunset is usually the starting point for the activity of nocturnal animals. Their activity generally stops some hours after sunrise, when the environmental temperature reaches a specific threshold. This, in its

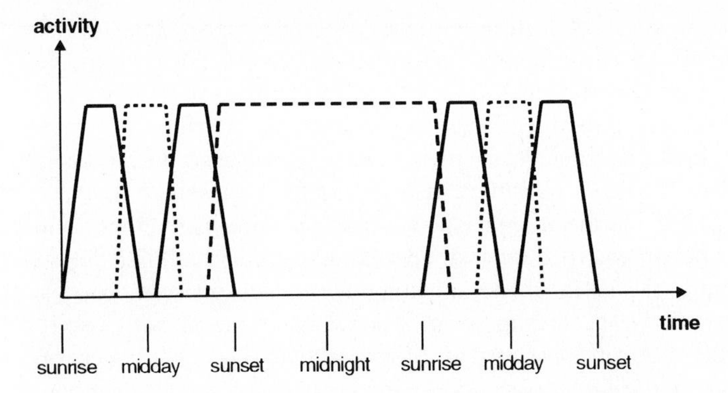

Fig. 22. The entire 24-h period is exploited by arid-adapted animals. *Continuous line* Diurnal bimodal animals, such as the tenebrionid beetle *Pimelia grossa* during summer (Sect. 4.2); *dotted line* diurnal unimodal animals such as the anthicid beetle *Anthicus fenestratus* (Sect. 4.4.4); *dashed line* nocturnal animals

turn, stimulates the activity of diurnal species. This is only a schematic description of what happens daily in a desert ecosystem, since, in fact, there are more than two possible temporal niches. For example, several animals are crepuscular; that is, they are active only in the morning and in the evening. Others are endowed with different kinds of bimodal activity (Sect. 4.2); and so on. All hours of the day and night are exploited by different arid-adapted animals, every one of which may avail itself of species-specific behavioural patterns to ensure optimum conditions. Some animals even occupy the most severe temporal niches so that food competition and/or predation are minimized.

Not all animal activity is carried out above ground. Many fossorial animals spend much of the time in their holes consuming stored food; many animals mate sheltered from indiscreet (or interested!) eyes; and numerous social species are very active underground. The prolonged and apparent inactivity of predatory desert animals may really mask patient waiting for prey. This is typical of snakes. Sometimes, however, it may be a matter of climatic adjustment, which is the case with sun-basking and fog-basking animals (Sects. 4.4.1 and 4.4.5).

4.2 Modulation of Activity Rhythms

Routine stereotyped behaviour is timed according to periodicities, which are regulated by biological 'clocks' synchronized by environmental factors (*Zeitgeber*) (Cloudsley-Thompson 1980). Most rhythmic activities follow a periodicity of approximatively 24 h (based on a so-called circadian clock) and utilize the solar photoperiod as Zeitgeber. Others follow a periodicity of approximatively 1 month, based on a so-called circalunar clock, since the moon cycle is 29.5 days. For marine animals, it is frequently called a 'tidal clock' since tidal rhythmicity is directly affected by the attraction exerted by the moon on the sea. However, tides reach their lowest point twice during each lunar month, so that the tidal rhythm is also called the 'semilunar rhythm' (Saunders 1976). Many marine animal species have a circalunar clock, but various biological phenomena of terrestrial species also show this kind of periodicity. For example, the larva of the ant lion *Myrmeleon obscurus* carries out its pit-building activity according to an exact lunar rhythm, with maximum activity at full moon (Youthed and Moran 1969). The emergence of the African mayfly *Povilla adusta* occurs on the second day after full moon (Hartland-Rowe 1955, 1958). Littoral animals of sandy shores, such as sandhoppers (Papi and Pardi 1954, 1959, 1963; Papi 1960; Enright 1961) and carabid beetles (Costa et al. 1983a), which have a daytime solar orientation, are also able to orient themselves by night according to the lunar azimuth: the angle of orientation with the moon is regulated by a circalunar rhythm, independent of the solar rhythm. The menstrual cycle of women also follows a circalunar cadence. Other behavioural mechanisms occurring seasonally, for example migration and reproduction, are often based upon a 'circannual clock'.

Circadian, circalunar and circannual clocks are easily identified in the laboratory since, under constant conditions, their rhythmic activities take place with almost absolute regularity. Nevertheless, animals are not the prisoner of their endogenous clocks since they are able to adjust their activities to environmental events. Many desert animals can modulate their activity rhythms according to climatic conditions (see Bodenheimer 1934; Fiori 1956; Crovetti 1970; Mellini 1976a,b). They may cease activity to avoid climatic extremes, or prolong it to exploit particularly favourable conditions. All animals can manage to free themselves from their endogenous rhythm so that their rhythmic behaviour is plastic and adjusted to the actual environmental situation. This phenomenon may involve single individuals or an entire population: moreover, it may occur over single days or throughout an entire season. This variability in animal periodicities may make the observer think that exogenous rhythms are not involved. For example, the desert locust shows an apparent irregularity in its locomotory activity; but, it actually possesses an excellent endogenous clock, whose functioning is regulated according to environmental conditions (Cloudsley-Thompson 1977b).

Hamilton (1975) showed that some tenebrionid beetles, living in the Namib desert, are able to modulate their activity rhythm according to the soil-surface temperature: during the midday hours they cease moving so that their activity rhythm becomes bimodal. Many other arid-adapted tenebrionids employ the same self-protective strategy. Alicata et al. (1979) discovered that the Palaearctic species *Pimelia grossa* maintains two different periods of activity (morning and afternoon) in summer and only one (the central

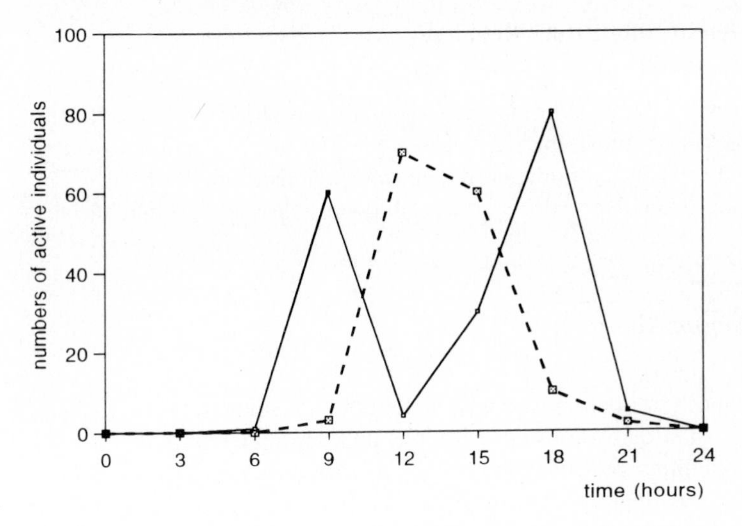

Fig. 23. Seasonal modulation of the activity rhythm of the Sicilian tenebrionid beetle *Pimelia grossa*. *Continuous line* Summer; *dashed line* Winter

part of the day) in winter (Fig. 23). Surface activity of the Negev desert *Hemilepistus reaumuri* is unimodal from February to March (the period of pair formation); it becomes bimodal from April to November (the period of gestation and growth) and ceases completely from November to January (the period of lowest burrow temperatures) (Shachak 1980). The Namibian *Lepidochora discoidalis* regulates its rhythm according to the wind (Louw and Hamilton 1972); the North American ant *Novomessor cockerelli* resorts to a complete inversion of its activity rhythm in summer, shifting from diurnal to nocturnal habits to avoid high daytime temperatures (Whitford and Ettershank 1975).

Vertebrates, for example, the ultrapsammophilous Namibian lizard *Aporosaura anchietae* (Louw and Holm 1972), may also modulate their temporal choice. The activity pattern of this diurnal species is typically bimodal, but when the temperature decreases too much or hot winds are blowing, it becomes unimodal. In an analogous way, the Asian rodent *Meriones hurrianae*, which is active at dawn in summer, shifts its activity to later in the day during winter, thus avoiding low crepuscular temperatures (Gosh 1975). Again, the North American desert lizard *Dipsosaurus dorsalis* shows unimodal surface activity (at midday) during autumn and winter, and bimodal surface activity (in the morning and in the evening) from April to September (Porter et al. 1973).

4.3 Exploitation of Substratum Resources

Animals may solve their thermohygric regulative problems in desert ecosystems by exploiting substratum features. They can utilize natural cover or penetrate the soil layers for longer or shorter periods; sometimes they are able to discover invaluable water resources hidden below the ground surface.

4.3.1 Subterranean Life

Many desert and semi-desert animals, belonging to several taxa, maintain a more or less stable relationship with the underground environment. The soil structure favours the establishment of thermohygric gradients between the surface and the underlying layers, so that fossorial species can find an optimum regardless of the external conditions. Moreover, these gradients, being subject to day-night inversions, regulate the vertical movements of the fauna and inform the buried animals of the climatic situation above ground.

Species living in permanent contact with the soil form the subterranean fauna; in arid areas, according to Wallwork (1982), this includes various groups of invertebrates (earthworms, nematodes, enchytraeids, geophilomorph centipedes, springtails, thysanurans, thrips, pseudoscorpions, and mites) and of small mammals (marsupial moles, talpids, and mole rats). Subterranean animals, like cavernicolous animals, spend their life without ever seeing the sunlight, thus showing an evolutionary trend toward the reduction in body pigmentation and eyes. Subterranean mammals also have a tendency for hairlessness. This reaches its maximum expression in the naked mole rat *Heterocephalus glaber*, which lives in the arid regions of Kenya, Ethiopia and Somalia (Hill et al. 1957). Unlike troglobites, fossorial animals have acquired morphological adaptations for digging and moving in the soil, but they have not lost their circadian and seasonal rhythmicities, thanks to indirect but continuing information, through thermohygric gradients, on what is happening externally.

4.3.2 Digging Behaviour

Several surface-active (epigeal) animals, namely species of relatively small body size, can dig and build nests or holes of suitable shape and depth. In the sandy substratum, burrowing habits are particularly widespread: the composition of the sand makes excavation easy, provided adequate morphological features and behavioural patterns are present. Burrowers are represented by both invertebrates (wood lice, arachnids, centipedes, millipedes, insects) and vertebrates (amphibians, reptiles, mammals). They spend part of the day and/or year underground, thus avoiding both thermal extremes and drought. It has been calculated that in the Karakum desert the thermal midday difference between the soil surface and 10 cm below is about 17 °C (Leopold 1962). In the Mojave desert, while the sandy surface reaches 65 °C, the temperature may be only 16 °C at a depth of 45 cm (Leopold 1962). Moreover, underground moisture remains rather high, so the animals can minimize their water loss.

4.3.3 Looking for Sheltered Microhabitats

Various arid-adapted animals occasionally or regularly utilize the burrows of other animals. For example, centipedes occupy the abandoned holes of scorpions and small mammals (Wallwork 1982). The large millipede *Orthoporus ornatus* retreats into the nests of harvester ants during the dry season (Wooten and Crawford 1975). The burrowing owl *Speotito cunicularia* utilizes holes made by mammals.

Other species take refuge in sheltered microhabitats: in crevices under rocks or stones (for example, scolopendromorph centipedes, scorpions, uropygids and lizards), in the organic litter under bushes (spiders, uropygids and snails), and so on. Many animals, mostly insects, excavate holes in trees and cacti. The above-mentioned gila woodpecker gouges cavities in the stems of saguaro cacti looking for insect larvae: these cavities can be used by numerous hole-nesting birds (e.g. elf owls, screech owls, purple martins, sparrow hawks, gilded flickers).

Some psammophilous animals do not dig holes, but merely penetrate into the sand. This often occurs among tenebrionid beetles, sand roaches, lizards, snakes, marsupial and golden moles. For example, the North American horned lizard *Phrynosoma cornutum* can disappear quickly head first into the sand, wriggling its body from side to side. When the soil temperature becomes too high, it emerges, looking for shaded areas (Sutton and Sutton 1966). Other lizards, such as skinks (Anderson 1898; Lawrence 1959), lacertids (Lawrence 1959), iguanids (Stebbins 1944; Norris 1958) and agamids (Schmidt and Inger 1957; Pope 1960), bury themselves in the sand.

Some sand-dwelling grasshoppers are also able to bury themselves: at dusk various species of the genus *Acrotylus* cover themselves up with the sand and remain motionless during the night and early morning (Fig. 24); when the surface temperature of the ground is warm enough, they become active (Nagy 1959; Conti et al. 1995). Also some snakes, such as the sidewinder rattlesnake *Crotalus cerastes* (Klauber 1956), can pull sand over themselves.

Fig. 24. The Palaearctic grasshopper *Acrotylus longipes*. *Above* An individual during self-burial; *below* finally, only eyes and antennae are visible

4.3.4 Exploitation of Water Resources

Various behavioural adaptations of desert animals enable them to exploit water resources. Water is a fundamental element for both hygric regulation (Fig. 25) and cooling the body (Fig. 26).

It has been known for a long time that Namibian zebras are capable of locating underground water pools. When moisture has been perceived with the nose, they dig a deep hole with their hooves until they reach the water. Due to this extraordinary ability, the German geologist, Henno Martin, called zebras the 'hydraulic engineers of Namib'! More recently, Hamilton et al. (1977) discovered that during the driest periods of the year the Namibian gemsbok excavates sand in the dry bed of the Kuiseb river to construct semi-permanent water holes.

Another interesting behavioural pattern is exhibited by Namibian beetles of the genus *Lepidochora* (Seely and Hamilton 1976). These flying-saucer-shaped tenebrionids construct narrow trenches on the dune slip face perpendicular to the direction of the fog-bearing wind. The ridges of these trenches facilitate the collection and condensation of fog. When the beetles return along the trenches, they drink water from the ridges of sand.

Fig. 25. A black rhinoceros cow (*Diceros bicornis*) and her calf beside a water pool at night

Fig. 26. Adults and young of the African elephant wallow in a water pool

4.4 Exploitation of Body Resources

Desert animals can directly exploit body peculiarities to guarantee their thermohygric regulation. That is, they have access to personal resources to increase or diminish heat storage or to obtain the hygric support necessary in an environment substantially lacking in water.

4.4.1 Sun-Basking

A simple thermoregulatory behaviour pattern is sun-basking. It is frequently seen in many arid-adapted species and is the more or less prolonged motionless exposure to sunlight. During this apparent inactivity, the animals store sufficient heat to begin locomotor activity after the cool of the night, or to face a forthcoming nocturnal temperature drop.

When they emerge from their nocturnal shelters, many reptiles first expose only their heads to the sun's rays for several minutes before completely emerging (Mayhew 1968). According to Stebbins (1954), such behaviour is

Fig. 27. An emperor penguin basks in the sun

necessary to ensure sufficient warming of the central nervous system and the consequent ability to move quickly if necessary.

Sun-basking is widespread among ectothermic animals, and is also observed in some homeotherms. For example, the North American roadrunner spends the first part of the morning exposing its back to the sun. During this period of inactivity, the bird raises the feathers of its back and neck, using its highly vascularized skin as a 'solar panel' (Calder and Schmidt-Nielsen 1967). Moreover, in the polar regions, sun-basking is essential for thermoregulation (Fig. 27).

4.4.2 Self-Production of Shadow

Shade is invaluable to desert species. When the temperature becomes unbearable, most diurnal animals look for shaded places. There is a time limit for this. Unfortunately, vegetable cover is scarce in arid regions, thus, continual efforts are required to find shade. Numerous animals of different species gather under scattered bushes and, therefore, have to face the various problems of living together in limited areas. Moreover, in such promiscuity, the risk of cannibalism and predation is greatly increased.

An ingenious alternative solution is to make one's own shadow. The African ground squirrel *Xerus inauris* raises and flares its bushy tail casting a shadow above its body which is oriented away from the sun (Marsh et al. 1978). Some ground-nesting desert birds, such as the ostrich and the sand grouse, shade their eggs lying on the soil surface with their bodies. Again, the sand grouse, raising its body and opening its wings above the nest, provides its eggs with shade and ventilation (Dixon and Louw 1978).

4.4.3 Body Orientation

Orientation of the body to the sun is another important thermoregulative mechanism, not confined to desert animals (Louw and Seely 1982). Fraenkel (1930) and Kennedy (1939) observed that thermophilous insects, such as the desert locust *Schistocerca gregaria*, expose the long axis of their bodies perpendicularly to the sun rays, thus maximizing heat gain. Kennedy (1939) demonstrated that, in this spatial configuration, grasshoppers expose the maximum body surface to the sun. In contrast, I discovered that some homochromous acridids (*Sphingonotus candidus personatus* and *Acrotylus longipes*) on sandy beaches of Sicily orient the long axis of their body according to the direction of the sun's rays. This behavioural pattern may represent a suitable adaptation to overcome the hottest hours of the day. However, since this behaviour occurs throughout the day, I presumed that its primary function must be to increase concealment, for anti-predatory purposes in a sandy, vegetationless environment by minimizing the shadow on the ground (Costa 1970a, b).

In desert animals, orientation of the body in relation to incident solar radiation is widespread. It may be found in both invertebrates, for example the Namib long-legged beetle *Stenocara phalangium* (Henwood 1975), whose global thermal strategy will be described in the Section 4.4.4, and the North American grasshopper *Taeniopoda eques* (Whitman and Orsak 1985), and in vertebrates, for example the springbok (Hofmeyr and Louw 1987). The springbok orients the long axis of the body laterally or parallel to the solar beam according to air temperature. Other examples include lizards, such as the zebra-tailed lizard *Callisaurus draconoides* (Muth 1977; Cloudsley-Thompson 1991), ostriches (Louw et al. 1969), and various mammals, such as camels, zebras, wildebeest and species of antelope (Cloudsley-Thompson 1993; Louw 1993).

4.4.4 Locomotor Strategies

When the soil temperature soars, insects may rely on their long legs to increase as much as possible the space between the ventral surface of their body and the overheated ground below. For example, this occurs in the small beetle mentioned above (*Stenocara phalangium*) which inhabits flat interdunal valleys of the Namib desert. Early in the morning this insect keeps its abdomen close to the soil surface as it moves, while the long axis of the body is oriented perpendicularly to the sun's rays. As the surface temperature rises, however, it extends its very long legs and walks as though on stilts (Henwood 1975). A similar behavioural pattern is exhibited by the Palaearctic acridids *Acrotylus longipes* and *Oedipoda miniata* which live in Sicilian coastal dune systems (Alicata et al. 1982). However, grasshoppers may also rely on their capacity to jump when in search of shaded areas. Jumping suddenly breaks contact with the burning soil surface. Other animals may utilize different resources to avoid excessive substratum temperatures during periods of locomotion or rest. A particular locomotor pattern is exhibited by some desert snakes. These legless reptiles are restricted to crawling on the ground; but some snakes in sandy areas, such as the horned vipers of the Great Palaearctic desert, the Peringuey's adder of the Namib and the sidewinder, the rattlesnakes of the Great American desert, can move diagonally, leaving characteristic tracks on the sand surface (a series of separate S-shaped furrows) (Mosauer 1932a, b, 1933; Sect. 6.1.1). This winding locomotion facilitates movement on the shifting sand and reduces body contact with the hot soil. In addition, it enables the snake to approach prey unobtrusively (Sutton and Sutton 1966; Cloudsley-Thompson 1977a; Robinson and Hughes 1978).

A detailed example of complex and highly specialized thermoregulatory behaviour is furnished by the ultrapsammophilous Namib desert lizard *Aporosaura anchietae* (Louw and Holm 1972). As the surface temperature reaches about 30 °C, the animal leaves its cover beneath the sand and adopts a characteristic sun-basking posture, raising its four legs and tail. When its body is sufficiently warmed, it begins its locomotor activity on the dune slip faces. As the surface temperature reaches about 40 °C, however, the lizard raises its body as high as possible above the heated soil and periodically performs an extraordinary 'thermal dance' (Louw and Seely 1982; Seely 1987), raising alternatively its diagonally opposite limbs and utilizing its heavily cornified tail as third support for the body.

Another widespread thermoregulatory behaviour pattern appears in arid-adapted animals which carry on their activity during the hottest part of the day. It consists in running rapidly over the overheated substratum and is most frequent in species that are unable to dig. On some sandy beaches of Sicily, at midday it is easy to see a small anthicid beetle, *Anthicus fenestratus*, which runs so very fast on the sunny ground surface that it may even strike the observer as it runs or flies closely above the sand (Alicata et al. 1982).

Rapid runners have very long legs. In desert environments, representative examples are the tenebrionid beetles belonging to the mainly South African genus *Onymacris* (Penrith 1975) and the central Asian and North African genus *Adesmia* (Medvedev 1965).

4.4.5 Fog-Basking

The exploitation of fog, dew and water vapour is fundamental to desert plants and animals (Louw and Seely 1982). Fog is a constant characteristic of cool coastal deserts (see Sect. 2.3.2): they are therefore called 'fog deserts' (Meigs 1953). In such environments, various animals can make use of condensed fog by drinking the water droplets which have formed on the substratum or on the vegetation. For example, jackals living in the Namibian Skeleton Coast lick condensed fog from the rocks (Louw and Seely 1982); while the Namib lizard *Aporosaura anchietae*, mentioned above, drinks fog droplets that have condensed on the sand dunes (Robinson 1980).

Many desert animals can absorb water through their hygroscopic skin: they may expose their bodies to the air during periods of fog, thus maintaining an optimum water balance. In anurans, the cutaneous absorption of water is fundamental for water uptake (Chew 1961); but in various other animal groups also, the absorption of water from moist air frequently occurs. The best-known example of fog-basking is provided by the Namibian tenebrionid beetle *Onymacris unguicularis* (Hamilton and Seely 1976). This diurnal insect normally spends the night buried below the dune sand; however, when nocturnal fogs appear, it emerges and reaches the dune crest, where it adopts a particular posture. It puts its body in a head-down quasi-vertical stance with the dorsal surface exposed to the fog-bearing wind. Condensed fog droplets, gliding downward along the elytral grooves, reach the mouthparts of the insect, so allowing it to drink. A very similar head-down posture is adopted on foggy nights by another Namib tenebrionid beetle, *Physadesmia globosa*, whilst sucking moisture from damp sand. According to Cloudsley-Thompson (1990), this behaviour may provide an explanation to the evolution of fog-basking in *Onymacris unguicularis*. On the other hand, the ethogram of other desert coleopterans includes a convergent defensive behaviour, named 'headstanding', sometimes associated with the emission of repellent secretions (Crawford 1981).

4.5 Exploitation of Social Resources

In certain cases, desert-adapted animals may have access to social resources to solve their thermohygric problems. Eusocial insects (Isoptera and Hymenoptera), have specialized behavioural patterns which guarantee a highly stable microclimate in the nest. Ettershank (1971) showed that the mound built around its nest entrance by the meat ant *Iridomyrmex purpureus* can lower the temperature in the upper portion of its subterranean burrow system. Furthermore, termite mounds are always oriented spatially in such a way as to control the heating of air and evaporation in the nest. Termites often build their termitaria in protected places (Fig. 28). In the Australian species *Amitermes meridionalis*, an extraordinary mound is built, oriented in a north-south direction, possibly in response to magnetic cues (Grigg 1973). This orientation results, to some extent, in thermoregulation throughout the day.

The subterranean nest of the central Asian desert termite species *Acanthotermes ahngerianus* is constructed so as to achieve a suitable level of evaporation deep in the soil (Ghilarov 1960). Wood and Sands (1978) pointed out that, in hot desert soils, the subterranean nests of termites can affect the percolation, storage, and drainage of water, favouring also the growth of

Fig. 28. Termitarium close to vegetation

plant roots. In the desert isopod *Hemilepistus reaumuri* also, microclimatic regulation is favoured by social organization (Shachak 1980; Sect. 9.1).

Social behaviour may also solve thermal problems in vertebrate animals. This is the case among Antarctic penguins. For example, the emperor penguin *Aptenodytes forsteri* (Prevost 1961) adopts a collective strategy in the face of the terrible polar winds which may blow at over 100 km/h, causing the temperature to drop below −100 °C! During the brooding period, the males draw together, beside one another, with their backs turned outward, forming a barrier against the mortal cold. Individuals on the periphery shift periodically toward the centre of the colony. In this way, the penguins succeed in preventing their bodies, and those of their young, from freezing.

Among vertebrates inhabiting hot deserts, it is also possible to find social behaviour having a thermoregulative function. For example, dromedary camels crouch in groups, thus reducing the total body surface exposed to the sun's rays (Wilson 1990).

5 Self-Protective Mechanisms

All the behavioural patterns that preserve animals from environmental challenges should be included in this chapter. Many of these, however, have already been described in the preceding chapter. Therefore, only the ethological mechanisms which maximize protection from predators will be treated. This can be achieved by several strategies, in which morphological and behavioural features concur. Various expressions have been employed by students of animal behaviour to describe them. Not surprisingly, therefore, a certain confusion of terms exists. Expressions such mimicry, protective resemblance, camouflage, masking, homochromy, concealment, crypsis, adaptive colouration, protective colouration, etc. are not always used correctly (Cloudsley-Thompson 1986). Therefore, I will give my own opinion on this matter.

'Animal camouflage' (or masking) can be achieved by different techniques. Mimicry is one of them, which means 'imitation of a model' by an animal. Broadly speaking, the model can be an animal belonging to a different species, a vegetable, to an inanimate object; however, I prefer to limit the use of the term mimicry to cases of imitation of an animal model. All other situations involving possible 'mimicry' can more properly be included under the term 'protective resemblance'. This is defined as 'aggressive resemblance', when a predator resembles something which attracts the prey. Two main kinds of mimicry are traditionally distinguished: Batesian mimicry, according to which an animal imitates the appearance of a species disagreeable to predators, and Mullerian mimicry, when different distasteful species present a similar aspect. In the latter, it is extremely difficult to discriminate between the model and its imitators. Furthermore, it is often subjective and arbitrary to have to judge mimicry as being either Batesian or Mullerian.

Many animals present body colouration quite similar to the background colour of their environment (homochromy or cryptic colouration). This phenomenon can be found in every habitat. Moreover, many homochromous species can imitate the features of the substratum perfectly. In addition to the colour of their exposed body surfaces for example, sand-dwelling animals may show dense dotting on the upper body surface which simulates grains of sand. Gravel-dwelling animals also have spots, which exactly resemble the alternation of light and dark portions of the background. Leaf insects (Phasmidae) imitate foliar veins on their wings; some moths resting on tree trunks have wing stripes that resemble the bark stripes, and so on. In my opinion, it is correct to use the term 'crypsis' in these cases. Homochromy is not merely a morphological phenomenon, since various physiological and behavioural patterns are involved in this kind of camouflage (Fig. 29). First, homochromous animals nearly always have a daytime rhythm of activity, or else they have nocturnal habits but are not able to conceal themselves from predators during daylight and so must remain motionless until evening comes. On the other hand, particularly in dry regions, the stars and the moon light up the night environment to a surprising degree. Consequently, even homochromous animals may need to escape nocturnal predators (Cloudsley-Thompson 1977a). Some animals are also able to change their colouration after displacement from one place to another – for example, from a shaded area to a sunny area – by automatic rearrangement of the pigment granules in their skin (Sutton and Sutton 1966). Some Arctic birds and mammals may change their body colouration twice a year. They are white during snowy winter and brown during other periods of the year. Finally, many homochromous carnivores may, at the same time, be invisible to both their predators and their prey.

Aposematism is the opposite of crypsis. Accordingly, instead of concealing itself within the environmental background, an animal species on the contrary draws the attention of potential predators to itself by bright colours and other striking features, as well as by warning sounds. By using this strategy, widely employed by Mullerian mimics, a moderately poisonous or distasteful animal recalls a former incident to the memory of an experienced predator. Since aposematic animals displaying the same garish signal may belong to different species, this self-protective technique represents a particular kind of interspecific co-operation. It is explained as mutual altruism, which passes over the boundaries of the species. The showy morphological characteristics, high population density and exaggerated slowness of aposematic species reinforce the negative experiences of predators; but any predator must learn personally to avoid unpleasant prey!

According to several authors, Batesian mimicry, crypsis and aposematism, as well as 'anachoretic behaviour' (that is, the utilization of barely attainable habitats, such as steep places or deep soil layers) are the main mechanisms of primary self-protection against predators: in these four different situations, the probability of direct encounters with predators is minimized. In contrast, retreat, escape, deimatic behaviour (that is, the display of intimidating aspects or postures), thanatosis (or 'feigning death'), deflection of attack, and aggression against the predator ('counterattack') are considered to be the six main types of secondary self-protective mechanisms. In all these cases, the prey maximizes its chances of survival after encountering an enemy (Edmunds 1974). I do not entirely agree with this classification of anti-predatory strategies. Aposematism does not eliminate aggression by predators since every predator must have at least one experience before learning its lesson and later, if very hungry, might even eat other aposematic prey. Moreover, it is not always

Fig. 29. Examples of homochromous desert animals. *Above* Namibian lizard; *below* Sicilian grasshopper

possible to distinguish between Batesian and Mullerian mimicry (Darlington 1938; Cloudsley-Thompson 1986). Finally, other kinds of mimicry, such as 'Mertensian mimicry' (Wickler 1968; Sect. 5.1.2), 'aggressive mimicry' and 'intraspecific mimicry', cannot be included in the two preceding categories.

Nevertheless, for descriptive purposes, I will adopt the above-mentioned classification. At the same time, I must include social defensive strategies as particular cases of secondary self-protective behaviour.

5.1 Primary Self-Protection

5.1.1 Anachoretic Behaviour

Anachoretic behaviour is of two different kinds, permanent and temporary. The former category includes the behaviour of comparatively sedentary animals which never abandon their protected habitats. Typical examples are species that spend their entire life underground, such as earthworms, moles, mole rats, etc. In desert environments, the most typical permanent anachorete is the African naked mole rat whose body clearly shows that it never comes to the soil surface (Sect. 4.3.1).

Temporary anachoretes spend part of their life in the open air. For example, many sand-dwelling insects, having strictly subterranean larval phases, later live as epigeous imagines. Moreover, there are numerous cases of desert animals which have alternating periods of external activity and inactivity in protected places.

5.1.2 Batesian Mimicry

This kind of mimicry, not yet extensively investigated in desert animals, is wide-spread among insects such Lepidoptera, Hymenoptera and Diptera. Studies on African butterflies have shown how Batesian mimicry is often associated with the complex phenomenon of polymorphism. The palatable females of *Papilio dardanus*, a species of the African arid savannah, can assume characteristics of three different unpleasant lepidopterous species (Bonner 1980). An extreme case of mimic polymorphism is that of *Pseudacraea eurytus*, which is able to imitate no less than 33 different distasteful species (Carpenter 1949). In desert environments, some black bee flies (bombyliid dipterans) are regarded as Batesian mimics of black bees (Cloudsley-Thompson 1977a). Other examples are provided by syrphid flies imitating wasps (Edmunds 1974). Among the vertebrates, the inoffensive African savannah-dwelling aardwolf, which is an insect-eater, can be mistaken by large predators for the formidable striped hyena *Hyaena hyaena*. Among snakes, mimics are widespread. For example, the North American long-nosed snake *Rhinocheilus lecontei* is a mimic of the poisonous elapid *Micruroides euryxanthus*. The egg-eating *Dasypeltis scabra* closely resembles the poisonous viperids with which it shares its habitat, *Echis carinatus* in Egypt and *Bitis caudalis* in southwestern Africa (Gans 1961; Fitzsimons 1962; Rose 1962). There is animated controversy regarding the interpretation of the mimics of several highly venomous coral snakes. An animal that meets one of them does not have the possibility of later exploiting its negative experience since it almost invariably dies! What advantage, therefore, do the imitators have of such deadly poisoners? Wickler (1968) hypothesizes that, in this case, the models are the slightly venomous species, which are more numerous among the Elapidae. These non-lethal animals may be imitated by both innocuous snakes (Batesian mimics) and more dangerous snakes (Mertensian mimics). The advantage for the Mertensian mimics could be in avoiding the waste of poison on animals that cannot be used as food. However, this hypothesis, as yet not proved experimentally, does not meet agreement by all students of animal mimicry. On the one hand, a well-founded suspicion exists that many birds and mammals may inherit an innate dread of the 'coral snake pattern' (Greene and Pyburn 1973; Smith 1975; Ward 1979). They do not need individual experience, since they are already instinctively aware of danger. On the other hand, one cannot exclude the possibility that so-called Mertensian mimics may instead be 'aggressive mimics' of non-dangerous species, thereby attracting naive prey. In addition, Pough (1988) maintains that, apart from their venom, coral snakes are distasteful. Therefore, coral-snake patterns could be aposematic and non-venomous snakes could be mimics of the coral snakes. Some experimental studies on the responses of potential avian and mammalian predators to artificial models, as well as the analysis of parallel geographic variations in coral snakes and colubrids, appear to support this hypothesis (Gehlbach 1972; Smith 1975, 1977; Greene and McDiarmid 1981).

5.1.3 Crypsis

Crypsis is very frequent in arid and semi-arid habitats. In many cases, it is achieved by homochromy. A considerable number of invertebrate and vertebrate desert animals, showing light body colouration, become blurred on the substratum background. They have grey, sandy, yellow, reddish, or white upper surfaces exactly matching the prevailing habitat colour. In sandy deserts, sand-coloured animals include numerous psammophilous invertebrates (arachnids and insects) and vertebrates (anurans, lizards, snakes, owls, nightjars, sand grouses, larks, rodents, foxes, camelids, gazelles, pronghorns, etc.). One cannot exclude the possibility that the fundamental pale colouration of desert animals may be due to a phenomenon of generalized depigmentation in dry environments (the so-called Golger's rule), but this does not explain the exact correspondence that exists between body and substratum colour. For example, North American lizards of the genus *Holbrookia* show a differing homochromous colouration according to the sandy or gravel environment inhabited. Furthermore, recent knowledge of 'industrial melanism' in the British geometrid moth *Biston betularia* and other Lepidoptera (Kettlewell 1955a, 1973; Lees and Creed 1975) has dealt a decisive blow to doubts about the adaptive value of body colouration as a self-protective system.

As described above, crypsis may be represented by suitable patterns: dots, small spots, mottles, stripes, and so on, making desert animals inconspicuous against the background in which they live. At the same time, cryptic animals have a paler homogeneous colouration on their under surface which weakens the shadowing, which could otherwise expose them to predators. Their crypsis is, moreover, enhanced by long periods of immobility, as well as by particular body orientations (Cott 1940; Sect. 4.4.3).

Of the animals capable of changing their body colouration, especially well known are the chameleons. Among desert animals that can change colour, the most widely known is the North American horned lizard *Phrynosoma cornutum*, which is able to assume a body colouration that changes from grey to red according to the predominant background colour (Sutton and Sutton 1966). The homochromous species incapable of changing their body colour must be able to select an appropriate substratum on which to live (Kettlewell 1955b; Sargent 1966).

Body form may also add a substantial contribution to crypsis (homomorphism). Well-known examples include phasmatids and orthopterans. The Mexican stick insect *Diapheromera velii* is highly cryptic on shrubs of *Dalea scoparia* (Crawford 1981), the Argentine proscopiid grasshopper *Astroma quadrilobatum* on the South American cresote bush *Larrea cuneifolia* (Rhoades 1977), and so on. Many cryptic desert species dwell on perennial shrubs (Orians et al. 1977), since these simultaneously offer food and protection to their guests. Certain arid-adapted animals living in pebbly or gravelly areas are stone-shaped. For example, some stocky grey apterous grasshoppers, such the Fuegian *Bufonacris bruchii* (Fig. 30) and the Namibian *Lathicerus cimex*, are nearly indistinguishable from the pebbles that surround them (pers. observ.). In addition, they remain quite motionless even when they are approached. It is interesting that some succulent desert plants also present a similar type of homomorphism: the American living-stone cacti and the South African rock plants of the genus *Lithops* are the most representative examples of vegetable crypsis (Wiens 1982). Some body features may also combine to camouflage animals that live in arid and semi-arid environments. For example, the pronotal thorns of the Namibian nara cricket *Acanthoproctus diademata* (Fig. 31) make the animal virtually unidentifiable within the thorny shrubs of its host plant (pers. observ.).

Another technique of crypsis is based on concealment of the body contour. This can be achieved by particular patterns of colouration (disruptive

Fig. 30. The stone-shaped Fuegian grasshopper *Bufonacris bruchii* is nearly practically indistinguishable among the pebbles of its habitat

Fig. 31. The nara cricket *Acanthoproctus diademata* is hardly distinguishable within thorny shrubs of the nara plant (*Acanthosycios horridus*)

colouration) which succeed in breaking up the outline of the body (Mottram 1915). Disruption of the apparent continuity of the body surface with the help of optical (and even psychological) tricks may prevent or delay recognition of prey by predators (Cott 1940). Very common in both invertebrates and vertebrates, disruptive colouration finds its most representative examples in giraffes and zebras. These normally conspicuous animals become almost indistinguishable at dusk when they are most likely to be attacked, thanks to their dark and light patches or stripes.

5.1.4 Aposematism

Many desert animals utilize warning signals to discourage predators from attacking them. Most aposematic messages to potential predators are visual. They are based upon striking chromatic and morphological features. The most frequent colour is black; but white, red and yellow are also often found. They may be one colour or combined to guarantee that the aposematic animal stands out distinctly against its background. The black tenebrionid beetles were discussed earlier (Sect. 3.2.11). In arid environments, other insects, such grasshoppers, carabids, scarabs, wasps, bee flies, and various birds, such wheatears and ravens, are black and so very conspicuous against their background (Cloudsley-Thompson 1977a). In addition to its thermoregulative property, the dark colouration of so many desert animals could result from generalized patterns of Mullerian mimicry. On the other hand, even the best arid-adapted species would fail to survive for long if they were easily preyed upon !

The form of the body itself, or certain parts of it, combined with colouration and behavioural characteristics such as slowness, gregariousness, postural displays, etc. can play an aposematic role. Ungulates attract the attention of predators to their sharp horns. Antelopes with pale coats, such as impalas, hartebeests and gazelles, have dark horns, while other ungulates with dark coats, such wildebeests and musk oxen, have light horns with black tips. Skunks display a typical intimidating posture, showing off their white-plumed tails, before emitting a stinking anal-gland secretion.

Olfactory and acoustic signals are often utilized by inedible species to discourage potential predators. Uropygids deter attackers with repugnant substances (Sect. 3.2.8). Various beetles, many lepidopterans, some hymenopterans and hemipterans produce similar unpleasant odours (Rothschild 1961). Mustelids often combine a black and white striped coat with fetid substances secreted by their anal glands which signal their presence strikingly. Aposematic sound production is common in aculeate hymenopterans. The humming is a clear warning signal to many possible predators: it preludes a potentially painful sting. An interesting example of Mullerian sound mimicry between sexton beetles and hymenopterans of the genus *Bombus* has been described by Lane and Rothschild (1965). In this case, aposematic colouration, odour and sound concur in dissuading both diurnal and nocturnal predators from attacking this type of prey.

Aposematic animals sometimes send deceptive messages to potential predators. These signals, since they do not correspond to any real danger to the potential predators, are more properly called 'pseudoaposematic' and will be treated later (Sect. 5.2.3).

5.2 Secondary Self-Protection

5.2.1 Retreat

This behavioural response, characteristic of all temporarily anachoretic animals, is adopted especially by epigeal animals having burrowing habits. They spend most of the time sheltered in their holes, burrows, dens, lairs, or in the soil but, because of trophic, reproductive or social requirements, they are compelled to carry out some activity in the open air. The shorter the distance from their shelter, the easier retreat will be. Consequently, many animals do not venture far from their protection. In this connection, crickets are typical examples. At dusk, during the reproductive period, the male must emerge from its burrow and disclose its location to females by its loud calls. This is a very critical situation for the cricket because many predators are also endowed with a sharp sense of hearing and depend on sound emission from their prey to locate it in the dark! So, at the slightest rustling, the chirping male must be able instantly to break off its acoustic signal and rapidly re-enter its hole. I have often had the opportunity of watching such conflicting situations in the Sicilian sand cricket *Brachytrupes megacephalus* (Caltabiano et al. 1982a) and in the Namibian giant cricket *Brachytrupes membranaceus* (Costa et al. 1987). It is interesting to note that, during the trophic period, these crickets, both young and neoadults, must look for food near their holes. In this case, as they are silent, they can only be captured by animals able to locate prey in the dark without acoustic cues.

Quickly moving animals, such as rodents and lagomorphs, can venture further from their burrows, but when necessary, they must be able to remember the exact sites of their shelters and return to them at breakneck speed.

Some animals have a particular kind of retreat, i.e. 'portable protection'. This is the case with pulmonate gastropods and tortoises which withdraw inside their shells, and with wood lice and hedgehogs, which roll into a ball exposing their calcified cuticle or their prickles respectively to predators.

5.2.2 Escape

The most common reaction of an animal to impending danger is escape. This can be running, jumping, swimming or flying, according to the morphophysiological and behavioural characteristics of the species.

Many terrestrial animals flee to avoid predators. Spasmodic attempts occur between prey and predator to outdo each other: success depends mainly upon the greatest possible speed and endurance. Pronghorns, hartebeests, wildebeests and springboks are the fastest of all hoofed animals: they can race for a long time at speeds of 80 km/h. Ostriches can run for 15 min at 50 km/h and, if necessary, for a few seconds at up to 70 km/h. Zebras and giraffes are capable of running at speeds of 64 and 48 km/h respectively. Cheetahs, however, can move even more quickly. They attain speeds of 110 km/h for a few seconds (Walker 1968). Lions can race at a speed of 80 km/h for brief periods (Attenborough 1979); the maximum running speed of spotted hyenas is about 65 km/h, while African hunting dogs and American coyotes can maintain speeds of about 50 km/h for long distances.

Additional strategies are adopted by prey and predators to attain their respective expectations (Curio 1976). I shall limit myself for the time being to discussing the case of the prey. First, they must always be on the alert. A continuous state of vigilance is fundamental for perceiving a predator as soon as possible and so keeping a safe distance from it. Before this distance has been diminished dangerously, the potential prey has to escape. During flight, several animals (lizards, hares, rabbits, gazelles, antelopes, etc.) follow an irregular path. Sudden, unpredictable changes in direction – zigzags, twists, loopings, bounds, and so on – have been called ‘protean displays’ and serve to disconcert fast predators.

Similar defensive reactions are shown by animals capable of jumping or flying. In addition, among insects such as moths, orthopterans and mantids, another specialized type of defensive behaviour (‘flash behaviour’) has been described. These homochromous animals possess a striking colouration on the inferior surfaces of the wings, visible only while they are flying or jumping. This attracts the attention of predators, but after alighting the colour disappears. Some anurans also exploit analogous flash behaviour: they present bright colours on the inner surfaces of their hind legs which are visible only during jumps. Acoustic flash behaviour is exhibited by certain orthopterans, which produce crashing sounds while flying, but remain silent on the ground.

5.2.3 Deimatic Behaviour

An animal, which finds itself within the range of a predator, may display threatening postures or movements. The threat is real when an animal possesses weapons such as horns, teeth, fangs, hooves, claws, or stings (see Cloudsley-Thompson 1982). Threatening displays, which may be combined with obnoxious secretions and/or warning colouration, play an aposematic role. However not all potential prey are endowed with truly effective weapons. Various unarmed animals, when they can no longer seek safety in flight, resort to bluffing. That is, they cheat and confuse the predator by means of pseudoaposematic signals and behavioural 'tricks' (deimatic behaviour).

'False eyes' or eyespots are widespread in the animal kingdom. They are present on different parts of the wings or body of insects (mainly lepidopterans and homopterans), fishes, anurans, reptiles and birds. In a number of cases, particularly among Lepidoptera, the deimatic function of big 'false eyes' has been demonstrated experimentally (Blest 1957). Usually concealed, these deceptive signals are suddenly shown to the predator by opening the wings. As a result of surprise, the predator may let the prey escape. In other cases, small 'false eyes' are located on non-vital parts of the body and are constantly displayed: these have the function of deflecting attack (Ward 1979). The American sparrow hawk and the African dwarf owl possess 'occipital faces' of unknown significance.

Another form of deimatic behaviour is based on the presence of 'false heads'. Animals belonging to different taxa utilize this misleading strategy. Such is the case in 'bifrontal defense', according to which some butterflies, by displaying a 'false head' on the hind wings, succeed in disconcerting predators by appearing to take flight backwards. Thus, the potential predator may attack a non-vital part of the body (Cloudsley-Thompson 1982). A 'double head' is also exhibited by amphisbaenians as well as several snakes, such as the African sand boa *Eryx muelleri* and the worm snakes (*Typhlops* and *Leptotyphlops* spp.). These reptiles possess a tail, the tip of which is very similar in appearance to the head, and when it is raised it conceals the true head. If a predator seizes their tail, they may have a chance to counter attack and escape. The 'false heads' of snakes are not considered pseudoaposematic signals, since these animals are certainly not unarmed. They are actually typical cases of deflection of attack (Sect. 5.2.5).

An apparent increase in body size is a frequent deimatic strategy in both aquatic and terrestrial environments. Fishes (Tetrodontidae and Diodontidae), amphibians (frogs and toads belonging to several families), reptiles (such as chameleons, iguanid and agamid lizards), birds (e.g. parrots) and even certain mammals (e.g. baboons), resort to this deceptive behaviour. A striking example from the desert is the Australian frilled lizard *Chlamidosaurus kingii*, which suddenly expands its enormous gorgets and opens its mouth widely, assuming a terrifying aspect (Cott 1940).

5.2.4 Thanatosis

Several predators do not feed upon dead prey which could be unpleasant, or even toxic if putrescent. Consequently, feigning death can be an effective anti-predatory strategy. Spiders, insects (such as beetles, mantids, cockroaches, phasmids, butterflies, hymenopterans, etc.), amphibians, reptiles and mammals often resort to thanatosis. Instantaneous immobility is combined with a posture similar to that of a dead animal. Body and legs present typical rigor mortis; appendices are retracted, the body is rolled up, and the animal may lie on its back. Other behavioural patterns may also be involved. The East African flap-necked chameleon *Chamaeleo dilepis* darkens its body, whereas the northwestern American snake *Heterodon nasicus* lies on its back, opens its mouth widely, and its oral and cloacal openings assume the red-violet colour of putrescent meat and release a fetid smell, thus perfecting the imitation of carrion (Ward 1979).

5.2.5 Deflection of Attack

One anti-predatory strategy is based on offering the less vulnerable parts of the body to predators. In this manner, many animals are able to limit physical damage and succeed in escaping or counterattacking when predators are too close and timely flight is not possible. Potential prey may attract the attention of prospective predators to unimportant parts of the body, such the spotted edges of the wings in the case of Lepidoptera (small 'false eyes', Sect. 5.2.3), or to organs that can be regenerated, such the tails of some lizards. The North American zebra-tailed lizard *Callisaurus draconoides* shows off its striped tail, and, like other lizards with a strikingly coloured tail (pseudoaposematic signal), is prepared to offer this organ to predators by autotomy (Ward 1979). Several insects also resort to autotomy, abandoning a leg to the predator. Some aposematic snakes deflect the attention and attack of predators to their tails. In certain cases, the tail imitates the head (Sect. 5.2.3), and an attack against it allows the snake to launch a counterattack with its real head.

Another well-known defensive behaviour, based on deflection of attack, is adopted by some ground-nesting birds (e.g. *Charadrius* spp.). These birds, simulating wing damage, succeed in luring a predator from the vicinity of their brood. This so-called injury feigning (Jourdain 1936) is not a self-protective behaviour, however, but an altruistic attribute of parental care.

5.2.6 Counterattack

The last defense of an animal attacked by a predator may be to counterattack, using all available weapons. Wildebeests and oryxes try to gore lions and other predators in their more vulnerable parts. A giraffe may succeed by using its hooves in forcing a lion to flee. The polecat resorts to spraying its stinking anal secretion. A double-headed snake (see above) can poison its enemy. A cornered rat may leap and bite the head of its predator, and so on.

Prolonged fights sometimes occur. The predator may encounter continuous retaliation from a dangerous and belligerent prey, thus exhausting it before delivering a deadly attack. The possibility of prey escaping depends upon its effectiveness in counterattacking, or on its physical endurance. A spectacular and well-known example is furnished by the attacks and counterattacks between mongooses and venomous snakes, such as cobras and vipers (Mendelssohn et al. 1971; Curio 1976). The mongoose, by changing its position continuously and repeatedly calling forth the reaction of its prey by 'mock attacks', attempts to tire it and, at the same time, to avoid its bites. The mongoose usually succeeds in this provocative strategy and, in the end, kills the snake: however, sometimes the snake overwhelms its tormentor, especially when the latter is young and inexperienced.

5.2.7 Social Defense Strategies

Social organization offers great advantages. In arid and semi-arid environments, this is particularly evident (Sect. 9.1). Individual self-protection can also benefit from communal defense tactics. Aggregation itself can be a strong deterrent to attack by predators: number and closeness of individuals play a fundamental role in both aquatic (schools of fish) and terrestrial environments (swarms of insects, flocks of birds, various kinds of mammal groups such as herds of ungulates or troops of giraffes). For this reason, many animal species respond to a predator's approach by bunching. Individuals that stray from their group are readily seized; and a common predatory technique is based on attempts to separate individuals from their group (Curio 1976; Sect. 9.1).

Further behavioural patterns are employed by social species for antipredatory purposes. Particular 'group formations' which discourage a predator's attack and protect the young may be displayed. For example, musk oxen form an open circle with the young animals in the centre, while wildebeest migrate in single file (Sect. 6.2.3).

Life in gregarious groups allows alternation between vigilance and repose. The time spent in vigilance can be inversely proportional to group size, as has been demonstrated in ostriches (Bertram 1980). In eusocial species, some individuals specialize in guarding the group. This occurs among insects such as Hymenoptera and Isoptera, in wood lice of the genus *Hemilepistus*, and in rodent *Heterocephalus glaber*. In matriarchal groups of dwarf mongooses, subordinate males regularly alternate in vigilance behaviour, emitting a systematic song that continuously informs all their partners about the situation (Rasa 1985, 1989). A similar 'watchman's song' has been postulated by Wickler (1985) as being employed by bird groups. In an analogous manner, the Arabian desert babbler *Turdoides squamiceps* shows an alternation of vigilance activity, and different individuals take turns to relieve the sentry.

In this species, rank position appears to depend upon the time spent on guard (Zahavi 1974a, b).

Emission of alarm signals is common in social animals, mainly birds and mammals. The warning sounds have some general characteristics, making it difficult for predators to distinguish the individuals emitting them. They persist for only few seconds, have a relatively high frequency, and do not present any modulation of frequency or amplitude. These signals often characterize behavioural patterns such as bunching and mobbing. In certain cases, there may be a complex warning language. For example, the vervet monkey, by using different alarm calls, can inform members of its own species (and other animals also) of the presence of snakes, leopards, eagles, baboons or Man (Seyfarth and Cheney 1990). The dwarf mongoose transmits, by warning vocalizations, information as to the type of predator, its distance and elevation (Benyon and Rasa 1989). In an analogous way, the meerkat *Suricata suricatta* emits different alarm notes depending on the presence of aerial or terrestrial predators (Ewer 1963). Visual and olfactory signals can also be utilized as alarm techniques. In particular, alarm pheromones are frequently used by eusocial insects, mainly ants.

6 Patterns of Movement

Motility is one of the most characteristic features of animals. Modification of the body for movement in different media represents the main course pursued by biological evolution throughout the animal kingdom. Locomotion allows animals to solve many environmental problems, from self-protection and nutrition to reproduction and social interactions. Movement may be continuous or discontinuous, periodic or aperiodic, individual or collective, small- or large-range, slow or fast. It may involve several mechanisms of orientation, and represents the basis of dispersal and the colonization of new territories.

6.1 Locomotor Patterns

Numerous locomotor patterns have been described in different systematic groups and appropriate to the environmental medium. Several animals employ different patterns. For example, some of them, such as various insects and mammals, are able to walk and swim, others, such as winged insects and birds, to walk and fly, and so on. Despite the great variation in the pattern of propulsive movements, the basic mechanical principles are incorporated in Newton's three laws of motion (Gray 1968).

6.1.1 Crawling

Legless animals move on solid substrata by crawling. The survival of non-parasitic protozoans depends upon the presence of water, even a thin layer in the soil. Here, they make use of particular organelles such as flagella or cilia (e.g. Roth 1958; Fawcett 1959). Amoebae crawl by the extroversion and retraction of pseudopodia (amoeboid movement). This movement is achieved by protoplasmic contraction (Allen 1959, 1961, 1962). Turbellarian flatworms use two different locomotor mechanisms: 'sliding' and 'telescopic crawling' (Moseley 1877; Chapman 1950; Clark 1964). The former is achieved by vibrations of the cilia on the ventral body surface within a layer of mucus secreted by the cutaneous glands. The latter is due to contraction and relaxation of the longitudinal and circular muscular bundles which cause extension and shortening of the body. The terrestrial rotifers (class Bdelloidea) are able to crawl by looping (Barnes 1968). They adhere to the substratum with their foot, extend the body and then fix themselves to the substratum with their rostrum. They then release and protract the foot before the next extension of the body. The undulatory crawling of nematodes is produced by waves of muscular contractions facilitated by the hydraulic skeleton and very elastic cuticle (Wallace 1958, 1959; Gray and Lissmann 1964). Crawling is also found in other invertebrates, such as oligochaetes and pulmonate gastropods. In earthworms, rectilinear movements are produced by coordinate series of contractions of the circular and longitudinal musculature, while ventral bristles act as supporting points (Gray and Lissmann 1938). Among snails, locomotion is achieved by small waves of contractions of the foot muscles, which follow each other from behind to forward movement; a mucous secretion makes gliding on the substratum easier (Lissmann 1945a, b).

Among vertebrates, snakes, amphisbaenians and some lizards move by crawling. In legless reptiles, movement is due to muscular contractions, the scales function as supporting points. Their trajectory is often regularly sinusoidal, but in limited spaces snakes can utilize a 'concertina pattern' (Gray 1946). In this case, a train of close loops flows along the long axis of the body. A peculiar locomotor pattern, found in snakes of sandy deserts, is sidewinding. Here, movement of portions of the body alternates with portions which remain for a moment stationary on the substratum (Mosauer 1932a, b, 1933; Klauber 1944; Gray 1946). The body moves sideways in a direction different from that of the axis of the waves, leaving behind it a series of characteristic tracks (Sect. 4.4.4).

6.1.2 Walking

When animals walk on dry land, some legs are moved while others are stationary and support the body according to particular co-ordinated patterns (e.g. Hughes 1952; Manton 1952a, b, 1953; Gray 1961).

The evolution of arthropods, from forms with extensive metamerism and several appendages (one pair per segment), has followed a clear trend towards a reduction of both. Few but effective legs make very fast movements possible on the ground. Arachnids and insects, with respectively four and three pairs of legs, are the best-adapted invertebrates for colonizing dry land. Terrestrial vertebrates are typically tetrapod animals, even if certain groups employ a bipedal gait (birds, kangaroos, jerboas, kangaroo rats, springhaas, Man, etc.) and others are legless (snakes, amphisbaenians and some lizards). Since support of the body requires at least three supporting points, bipedal animals have to utilize further devices to keep their balance. For example, kangaroos use their tails as a third support; birds and Man have feet with large soles.

Diplopods are very slow. Their walking is based on the use of a large number of legs, and differs very little from crawling. The legs are inserted ventrally, and their movement involves particular co-ordinated patterns, causing the body to glide on the soil surface. On the other hand, the numerous and short legs of diplopods (two pairs of legs in each diplosegment) are specialized for pushing into crevices (Manton 1968). Chilopods (but not geophilomorphs) move faster. Their legs are long and inserted laterally. They enable these epigeal predators to achieve high speeds. In the wormlike geophilomorphs, on the other hand, the legs are slender and do not substantially contribute to propulsion. These slow hypogeal centipedes move inside the soil by means of expansion and contraction of the trunk of the body in the manner of earthworms (Manton 1952b, 1965, 1968).

Well known is the general ambulatory pattern of insects (Hughes 1952). Hexapods follow the 'double tripod' model (Fig. 32a), according to which three legs (fore- and hind leg of one side, middle leg of the other side) are in movement, while the other three act as supporting points. In the subsequent step, the use of the legs is reversed. Of course, this model can change if the legs have undergone adaptive modifications. Mantids walk with only four legs (Fig. 32c), and do not use their forelegs which are modified for seizing prey. This, however, does not reduce their locomotor efficiency. For example, the epigeal desert mantid *Eremiaphila* sp. can run quickly across rocky areas of the Sahara (Dekeyser and Derivot 1959). Other mantids, including bush dwellers such as the desert mantid *Blepharopsis mendica* (Cloudsley-Thompson and Chadwick 1964), and some stem-dwelling orthopterans, such the Palaearctic *Tropidopola cylindrica* (La Greca 1947), can climb with the aid of only four legs (Fig. 32b).

Among tetrapod vertebrates, ambulation occurs in a tetrapod pattern. Accordingly, the foreleg of one side and the hind leg of the other side tend to alternate with the other two legs in protraction and in the supporting phases ('diagonal synchrony'). In slow walking (Fig. 33a), there is actually a slight

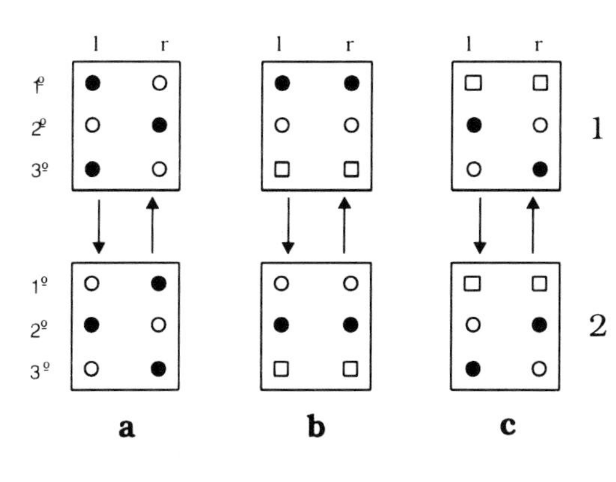

Fig. 32. Double tripod locomotor model of insects. *Symbols*: *1°*, *2°*, *3°* indicate the pairs of legs; *l* left legs; *r* right legs; *1, 2* subsequent phases of locomotion. *Circles* represent legs employed (*black* in movement; *white* supporting points); *squares* represent legs not involved. **a** General pattern; **b** *Tropidopola cylindrica*; **c** mantids

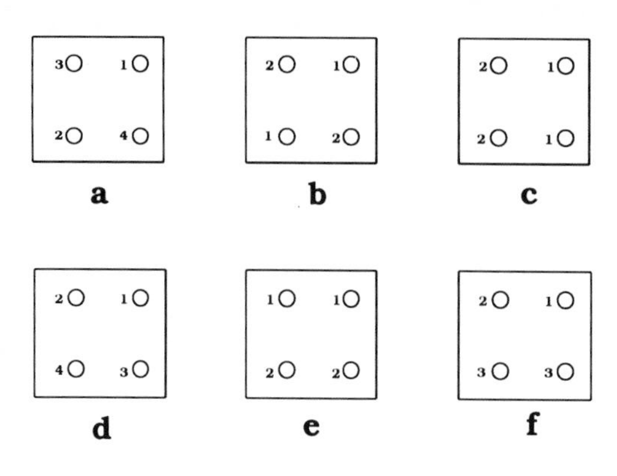

Fig. 33. Tetrapod locomotor model of vertebrates. *Circles* represent the legs employed, *numbers* refer to their sequence. **a** Walking (all quadrupeds); **b** trot; **c** rack (giraffes and camels); **d** transverse gallop (horses); **e** bound (squirrels); **f** half-bound (rabbits)

Fig. 34. Comparison between the trot of a springbok and the bipedal gait of an ostrich

phase difference in the employment of the legs to guarantee at least three supporting points continually. The diagonal co-ordinated model is observed exactly in the trot (Fig. 33b). In the gait of giraffes and camels, there is a 'homolateral synchrony' (Fig. 33c), while in the bound, characteristic of squirrels, the legs of each pair are moved simultaneously ('transversal synchrony') (Fig. 33e). In this case, the forelegs alternate with the hind legs in supporting and pushing the body, while, in the half-bound of rabbits (Fig. 33f), the two forelegs are moved with a slight shift. In the transverse gallop of horses, there is also a phase difference between the legs of the same pair (Fig. 33d). The bipedal gait of birds and Man is achieved mainly by means of a regular alternation of the legs (Fig. 34).

The jump requires synchronous use of the back pair of limbs; the third pair, in the case of orthopterans, the second in anurans, birds and mammals. Among kangaroos, kangaroo rats and springhaas, jumping is the only possible form of locomotion. Their hind legs and tail are very well-developed, while their forelegs are minute and unable to contribute to movement. Locomotion is nothing but a continuous series of jumps, realized by means of the hind legs with the help of the tail which acts both as a propulsive and a directional device. Jerboas have comparable morphological features and the same leaping habits, but they are also able to scramble up trees and shrubs. Climbing involves the incisor teeth, front legs and tail.

6.1.3 Digging

Many xeric animals exploit subterranean resources (Sect. 4.3.2). For this purpose, they must be able to dig into the ground. The basic digging organs of arthropods are the legs and mouthparts (Alicata et al. 1982; Wallwork 1982). The legs of fossorial animals, mainly the front legs, are shortened, thickened and flattened. In addition, they have digits and strong spines or claws. Digging occurs among wood lice, solifugids, scorpions, cockroaches, mole crickets, crickets, beetles, hymenopterans, spadefoot toads, tortoises, moles, and other vertebrates. Moreover, other structures, such as tufts of long hairs, may be used to scoop out and transport sand. Many rhaphidophorine gryllacridids possess tibial 'sand baskets' (Crawford 1981), while some wasps are equipped with tarsal-joint brushes (Pradhan 1957).

In many arenicolous insects, such as crickets, earwigs, carabid and scarab beetles, termites and ants, and in certain rodents, the head also contributes substantially to burrowing. In particular, the mandibles may be well developed and strengthened and may possess indentations. Some Asian voles, Middle Eastern mole rats, North American gophers and South American tuco-tucos use the incisor teeth as excavators, in addition to the forelimbs (Wallwork 1982). Various whip scorpions use their pedipalps to excavate burrows (Crawford and Cloudsley-Thompson 1971). Some Namibian and North American scorpions use their chelicerae in digging (S.C. Williams 1966; Newlands 1978). Some Saharan ants have 'psammophore organs', consisting of long curved hairs located on their head and employed to push sand away (Délye 1968).

A spectacular digging technique is performed by the Palaearctic sand cricket *Brachytrupes megacephalus* (Caltabiano et al. 1979b) and the Afrotropical giant cricket *Brachytrupes membranaceus* (Costa et al. 1987). The burrowing of these crickets involves various stages and implies, in addition

Fig. 35. Head digging by the giant cricket *Brachytrupes membranaceus*

to the use of legs, the active and significant use of the head (Fig. 35). The mandibles are repeatedly driven into the substratum vertically as shovels; the loosened sand is then carried backwards by the mandibles. The forelegs are kept close to the head, while the cricket moves backwards. From time to time the hind legs, thrusting powerfully backwards, are used to kick the sand away. When the tunnel is large enough to house the entire animal, the *Brachytrupes* turns around. It pushes the loosened sand back with its head and, just outside of the hole, throws it far away with a strong butt. The phases of sand removal by the mandibles, carrying the sand that has been removed backwards by the forelegs and mandibles, kicking it backwards with the hind legs and butting the accumulated sand away, are repeated many times until the tunnel has reached a suitable depth. Finally, the çricket obscures the entrance hole with some of the remaining sand, while a characteristic mound remains outside near the occluded entrance.

6.1.4 Swimming

Many terrestrial animals, both invertebrate and vertebrate, are able to swim. Locomotion in water is achieved by particular co-ordinated movements. These are very often different from ambulatory movement. At the same time, in a liquid environment, the legs are exonerated from the task of supporting the body. Even in desert and semi-desert environments, the ability to swim is

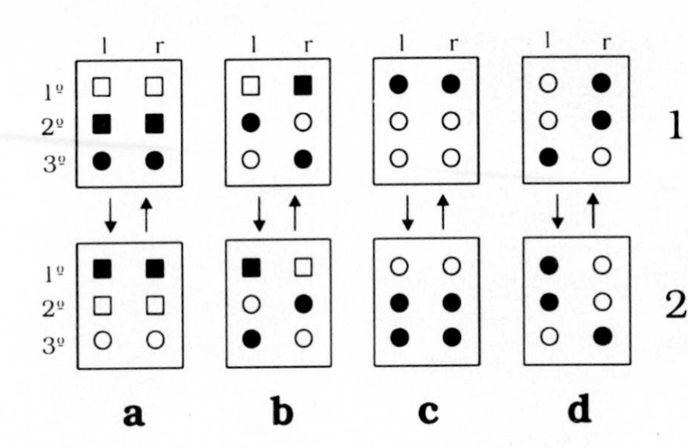

Fig. 36. Swimming patterns among some arid-adapted insects. *Black squares* represent legs which contribute only slightly to pushing the body; other *symbols* as in Fig. 32. **a** Grasshoppers; **b** mole crickets; **c** crickets; **d** carabid beetles

surprisingly widespread. For example, it has long been known to occur among hypogeal species such as mole crickets, moles and shrews (Selys-Longchamps 1862). The swimming pattern of the Sicilian cricket *Gryllotalpa quindecim* (Fig. 36b) has been described (Costa and Petralia 1979). It involves the active use of the second and third pair of legs to produce propulsion on the surface of water. These legs are moved alternatively backwards and forwards with a phase shift. So, when the middle and hind legs of one side are close together, the middle and hind legs of the other side are at thèir greatest distance apart. At the same time, the front part of the body (head plus pronotum), together with the forelegs, swings laterally to facilitate progression through the water.

Several epigeal arthropods, such as spiders, orthopterans, cockroaches (Fig. 38a), earwigs, mantids, carabid beetles, etc., are also able to float and swim.

In orthopterans, propulsion of the body through the water is essentially by the long and powerful hind legs. The general swimming pattern of Acrididae (Fig. 36a) is similar in both xerophilous and hygrophilous species. It consists of sudden backward movements of the hind legs, the forelegs are pushed forward near the head, while the middle legs are brought backwards close to the sides of the abdomen. The role of the second and third pair of legs is to propel the insect through the water, allowing it to assume an excellent hydrodynamic shape (Costa and Carveni 1975). Locusts also exploit the propulsive push of the hind legs and swim using the same motor pattern as in kicking (Pfluger and Burrows 1978). Various swimming patterns have been described in the Ensifera (Fig. 37). Some species exhibit a technique quite similar to that of the Acrididae, while others, namely crickets, use specialized co-ordinated movements, in which two or three pairs of legs are involved (Caltabiano et al. 1982b). A spectacular and very effective swimming pattern (Fig. 36c) is seen in the Palaearctic sand cricket *Brachytrupes megacephalus* (Caltabiano et al. 1979b). The legs of each pair produce synchronous propulsive strokes. The forelegs are stretched forward, and downwards, then extended laterally and backward. At the same time, the middle legs, extended laterally, shift to and fro in a horizontal plane, while the tibiae of the hind legs, previously

Fig. 37. *Gryllus bimaculatus* moving in a swimming mode

bent under their respective femurs, produce powerful backward strokes in a vertical plane. The same swimming pattern is shown by the Namibian cricket *Brachytrupes membranaceus*, which lives in the dry bed of the Kuiseb river (Costa and Petralia 1985).

Among earwigs, the ability to swim has also been found both in species living on sandy beaches, such as *Labidura riparia*, and species living far from water, such as *Forficula auricularia* (Caltabiano et al. 1979a). Swimming is performed by the three pairs of legs (Fig. 38b) accompanied by lateral swinging of the body. In the African savannah-dwelling mantid *Sphodromantis lineola*, Miller (1972) found a swimming pattern involving both the second and third pair of legs, which are moved alternatively. Caltabiano et al. (1981, 1983) described swimming patterns in mantids (Fig. 38c) and in some desert carabids (Fig. 36d). Well-known sand-swimming animals are the Saharan sandfish lizard *Scincus scincus* (Martens 1960), the Australian

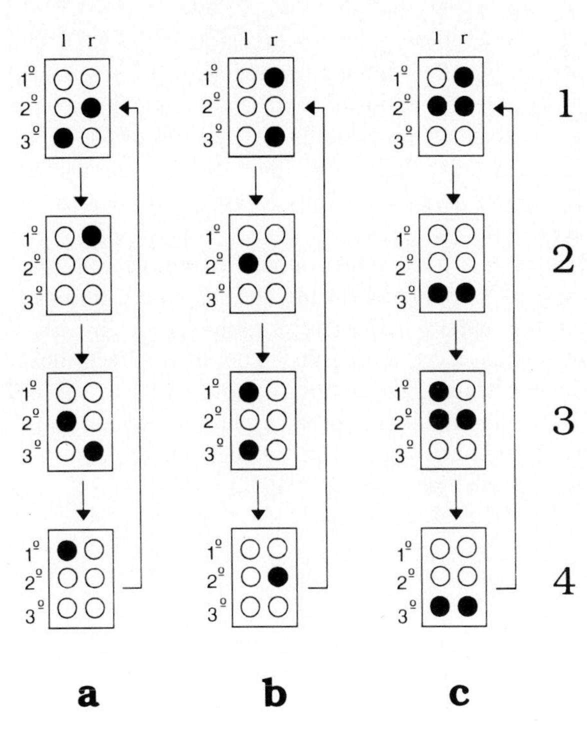

Fig. 38. Four subsequent phases of swimming patterns in **a** cockroaches; **b** earwigs; and **c** mantids. *Symbols* as in Fig. 32

marsupial mole *Notoryctes typhlops* (Walker 1968) and the Namibian golden mole *Eremitalpa granti namibensis* (Coineau 1981). However, I consider any animal that penetrates into the sand (Sect. 4.3.3) is, in essence, a sand-swimmer. This is seen, for instance, in various thysanurans (Watson and Irish 1988), tenebrionid beetles and sand roaches (Wallwork 1982), as well as in many reptiles, such as lizards and snakes (Pope 1960; Louw and Holm 1972; Bauer and Russel 1990; Cloudsley-Thompson 1991).

6.2 Orientation Mechanisms

Orientation plays an important role in animal behaviour. Animal movements are not usually random. On the contrary, they generally follow definite directions, whose choice provides the solution to various impelling problems. For this reason, many students of animal orientation, excluding primary orientation (control of suitable postures at all times), define oriented locomotion as 'goal orientation', or secondary orientation. Goals can be coincident with spatial cues (direct or immediate orientation). For example, this happens when an animal directs itself towards or away from an olfactory signal (positive or negative chemotaxis), towards or away from a light source (positive or negative phototaxis), towards a prey or away from a predator. If the goal differs from the cue, the orientation is called indirect or mediated. A convincing example of this is provided by animals endowed with astronomical orientation. In indirect orientation, the goal can be spatially defined. It consists of a particular, restricted site – for example, in homing (Sect. 6.2.2) – or it may be undefined, for example, when an animal attempts to escape from a stress situation.

The cue can be immobile (e.g. a fixed light source) or mobile (e.g. the sun, the moon, prey). The angle of orientation to the cue may vary, thus the animal's direction consequently changes if the cue is mobile (e.g. in heliotaxis) – or it is constant. In this case, an animal that must hold a rectilinear course must 'compensate' for the shift in the reference point (e.g. the sun or the moon). In solar orientation, this ability involves a chronometrical sense of time. The cue is the astral azimuth, which can be seen directly or else inferred from the plane of polarized light (Frisch 1949).

6.2.1 Primary Orientation

Primary or positional orientation depends mainly on the force of gravity. It is obvious that any animal must adopt suitable postures both in locomotion and at rest. At the same time, a well-balanced body position is indispensable for carrying out each behavioural programme. The ethogram of any species may therefore include various postural patterns (Fig. 39). For example, when browsing, a giraffe must stretch its head upwards to reach the highest acacia leaves. However, in order to drink, it must spread its very long front legs and lower its head, sometimes even below the soil level (Fig. 40). Other postures are, of course, employed by the giraffe during walking, running, mating, defense against lions, etc.

Other positional patterns are not strictly dependent on gravity. We have already considered various possible situations among desert and semi-desert animals. Examples include the self-production of shadow by *Xerus inauris* (Sect. 4.4.2), body orientation to the sun for thermoregulative or cryptic purposes (Sects. 4.4.3 and 5.1.3), the fog-basking posture of *Onymacris unguicularis* (Sect. 4.4.5), deimatic displays (Sect. 5.2.3), thanatosis (Sect. 5.2.4), and so on.

6.2.2 Secondary Orientation

Oriented movements of the body allow access to places that fulfill important biological necessities ('positive goals', e.g. feeding sites, mating sites, etc.) or escape from places having negative features ('negative goals', e.g.

Fig. 39. The ground squirrel *Xerus inauris* in different postures. *Above* Just out of the burrow; *middle* on the alert; *below* walking

Fig. 40. Different postural patterns of the giraffe (*Giraffa camelopardalis*): one individual is on guard, while another drinks

unfavourable areas having extreme climatic conditions or numerous predators). A positive goal can also be another animal, such as a prey, a sexual or a social partner: in an analogous way, a negative goal can even be the predator itself. Excluding immediate orientation, based merely on continuous sensory contact by the animal with its goal, goal orientation requires the use of directional information as cues. Within a familiar territory, an animal may rely on known landmarks (visual or other types). This seemingly simple orientational mechanism, called 'pilotage', actually requires profound knowledge – that is, the possession of an acquired topographic map of the habitat (Papi 1992). The use of landmarks is frequent among vertebrates which move over familiar areas. It has also been found in some invertebrates, such as the Saharan desert ant *Cataglyphis bicolor* (Wehner 1972). Different orientational mechanisms are involved in the exploration of unfamiliar territory. They may be based on a random or systematic goal orientation. Random wandering has been described by Fraenkel and Gunn (1940) as 'kinesis'; it is typically exemplified by the behaviour of the wood louse *Porcellio scaber*, which frantically follows a tangled route when it reaches a dry place and stops wandering only when it has by chance reached a moist locality. Systematic orientation may be achieved through different patterns of movement (Jander 1975). 'Transecting' is movement in a straight line in any direction.

This is a method often employed by insects living in monotonous habitats that lack conspicuous landmarks, e.g. wide sandy beaches and expanses of desert. Many arthropods resort to yet other systematic search programmes. For instance, the desert isopod *Hemilepistus reaumuri* (Hoffmann 1983a, b) and the desert ant *Cataglyphis fortis* (Wehner and Wehner 1986) utilize a search pattern consisting of a number of loops of ever-increasing size.

The ability to use sun-compass orientation has been found in numerous animals belonging to many different groups and living in all kinds of environments. In true desert areas, well-studied examples are the desert ant *Cataglyphis bicolor* (Wehner 1972) and the desert lizard *Uma notata* (Adler and Philips 1985). Many arthropods living in sandy beaches, such as lycosid spiders (Papi and Tongiorgi 1963), amphipods (Pardi and Papi 1952; Pardi 1960), isopods (Pardi 1954), carabid beetles (Figs. 41, 42a; Costa et al. 1982), etc., also have the power of solar orientation. Moreover, the moon may be an important orientational cue (Fig. 42b) for many animals, including arid-adapted animals (Sect. 4.2). Wind is, of course, a fundamental factor

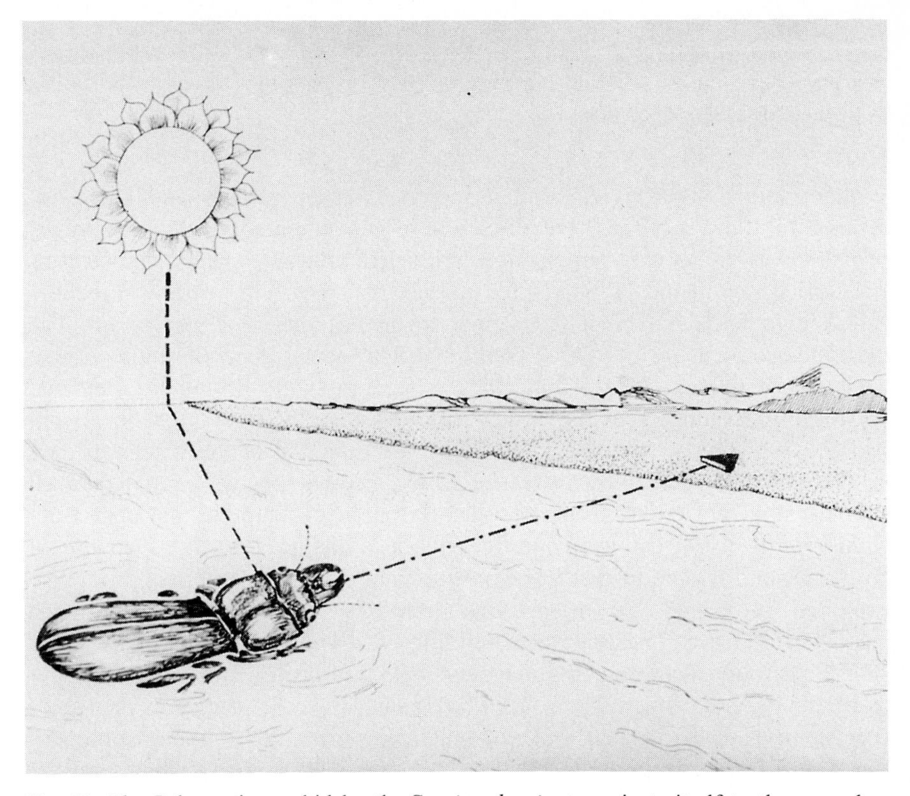

Fig. 41. The Palaearctic carabid beetle *Scarites laevigatus* orients itself to the sun when landing

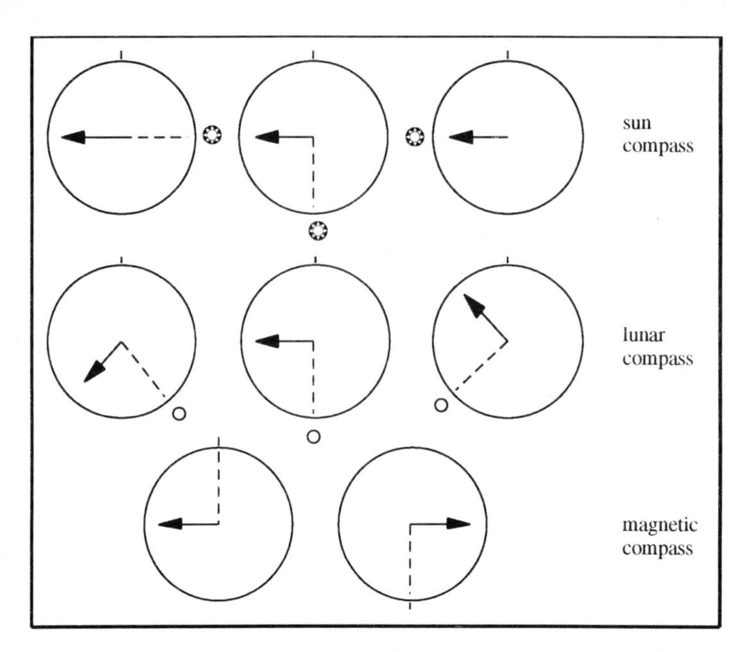

Fig. 42. Comparison among orientation patterns of *Scarites laevigatus*: ☼ the sun's azimuth; O the moon's azimuth; the *external dash* indicates the magnetic North; the *arrow* indicates the escape direction

in the orientation of flying animals, such as insects, birds and bats. According to Rainey (1976), the direction of movement of the desert locust *Schistocerca gregaria* is largely or wholly determined by wind. The swarms are carried downwind, although the insects may be flying in any direction, yet all together; they orient themselves to one another, not the ground. The radar studies of Schaefer (1976) confirmed that locusts generally move downwind, while Saharan pierid butterflies move crosswind during the day and Sudanese grasshoppers of the genus *Aiolopus* head SSW at night in all but strongly opposing winds. Whilst in flight, the direction of the wind can only be determined visually, by reference to the ground below. Even terrestrial animals may rely on wind as an orienting cue. For example, the desert ant *Cataglyphis bicolor* can use wind direction as a cue in orientation. An animal on the ground is able to feel the wind blowing. Magnetic information is also exploited by animals for orientational purposes. I have already mentioned magnetic influences on the mound building of *Amitermes meridionalis* (Sect. 4.5). Sensitivity to the Earth's magnetic field has been demonstrated in several animals, both invertebrate and vertebrate. However, a certain degree of skepticism as to the existence of a magnetic compass exists among some students of animal orientation (Able 1980). Nevertheless, recent studies on the non-visual orientation of littoral arthropods, living at different latitudes, have fully demonstrated its existence. In fact, both African equatorial sandhoppers

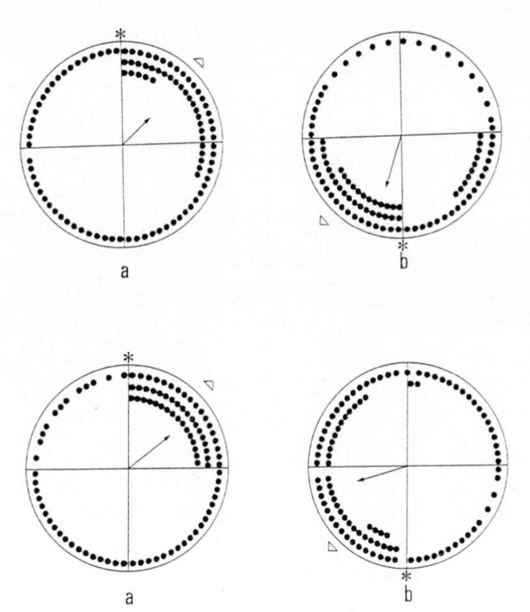

Fig. 43. Magnetic orientation in psammohalophilous insects: *above* the Namibian tenebrionid beetle *Pachyphaleria capensis*; *below* the Fuegian earwig *Esphalmenus rostratus*. *Asterisks* indicate the magnetic North; *external triangles* the expected mean direction; *internal arrows* the mean direction actually taken; *black circles* the animals: **a** Earth's magnetic field; **b** artificial magnetic field

(Pardi et al. 1984) and Mediterranean (Fig. 42c), Namibian (Fig. 43a) and Fuegian (Fig. 43b) insects (Conti 1994) can choose their directions of escape according to magnetic cues. Chemosensory and acoustic cues may also guide animals toward their alimentary, reproductive or social goals. Odours and sounds often initiate basic orienting reactions or taxes (immediate orientation). For example, positive chemotaxis leads male moths toward virgin females (Schneider 1966). Similarly, an irresistible positive phonotaxis leads female crickets toward singing males (e.g. Popov and Shuvalov 1977; Caltabiano et al. 1982a; Costa et al. 1987). Moreover, self-communicative techniques can be used for orientation. Examples are the scent trails followed by snails (Rollo and Wellington 1981), ants (Brun 1914) and other animals, and the echolocation of bats (Simmons et al. 1979). Thermal information may also direct some animals toward their goals. For instance, snakes such as boas and pit vipers are able to detect warm-blooded prey through the emitted infrared radiation (Bullock and Barrett 1968).

Secondary orientation plays a part in complex behavioural performances. The ability of animals to return to a familiar place after a more or less prolonged absence is defined as 'homing'. The 'home' may be an occasional shelter, the usual den or nest, the mating site, or even the birthplace. According to a scheme proposed by Griffin (1952) with reference to birds, and later extended by various authors to other animal groups, there are three main types of homing phenomena, corresponding to three different levels of ability: pilotage (if necessary, preceded by a random or systematic search for familiar landmarks); compass orientation (the direction is rigidly fixed); and navigation. The latter represents the most developed of animal capabilities for orientation because it allows an individual at any time to select the most suitable direction toward its goal. Further up-to-date classifications of homing phenomena have since been proposed (e.g. Jander 1975; Papi 1990). However, these will not be discussed in detail since they fall outside the scope of this book.

Navigation (a term sometimes used to indicate any phenomenon with indirect orientation) should actually refer only to charting a course to a remote goal that is beyond the limits of direct sensory perception (Schone 1984). It is not limited to homing, but may also play an important role in migration.

6.2.3 Dispersal and Migration

Under natural conditions an animal is born in a particular place within the habitat of the population to which its parents belong. On dry land, it is quite impossible for an individual to spend its entire life in its birthplace. In fact, the normal activity of an animal requires more or less continuous displacement in search for food, in escaping climatic extremes and other unfavourable environmental conditions, or escaping predators. Animals must also search for a sexual partner and secure protection for their offspring. The space utilized by an animal in this way may be defined as its 'individual habitat'. Borrowing a concept from the mathematical theory of sets, it is possible propose that the 'union' of the habitats of all individuals belonging to a population composes the 'population habitat'. The latter is not a spatially static unit because it may change from generation to generation in a dynamic and rather unforeseeable way. Individual movements within the existing population habitat do not influence its boundaries. Displacements beyond these spatial limits, however, modify the pre-existing population habitat. Even a casual departure can represent an event of dispersal and, at the same time, the first potential step in the colonization of a new area. Departure from the population habitat is termed 'emigration', while entry into the population habitat is termed 'immigration'. A sudden increase in population density is often the main stimulus to emigrate. Well-known examples of emigrations, caused by overpopulation, are those of the Scandinavian lemmings (see Curry-Lindahl 1962) and North African locusts (see Cheke 1978). Nevertheless, dispersal of certain species seems to occur regardless of current environmental stresses such as overcrowding. According to Howard (1960), there is also a tendency for genetic dispersal, which may manifest itself before saturation of the carrying capacity of the population and, for this reason, was defined by Lidicker (1975) as 'presaturation dispersal'. Such preventive behaviour appears to be widespread in rodents, such as mice, rats, voles, etc., and in other animals having a strong exploration tendency. An interesting example of apparent presaturation emigration over hundreds of kilometres was found in the central Australian desert rat *Rattus villosissimus* (Newsome and Corbett 1975). In an analogous way, the Mojave desert pocket mouse *Perognathus formosus* may disperse regardless of population density (French et al. 1975).

Many scientists (e.g. Dingle 1980) use the terms dispersal and migration interchangeably. Other students of animal behaviour (e.g. Heape 1931; Orr 1970) do not agree with such definitions and emphasize the distinction between migration and other movements such as emigration, nomadism, and passive dispersal. In fact, migration involves mass movements toward a definite, common spatial unit. It requires the use of precise mechanisms of orientation and occurs along common or statistically similar routes. Moreover, migration takes place regularly during a well-defined period of the year. Finally, various authors (e.g. Landsborough Thompson 1926) define 'true migration' as including return movements. So, true migratory species move to and fro (to-and-fro migration) or follow a subcircular route (loop or circular migration) between at least two different habitats. Return migration is actually an example of homing on a large scale. The heterogeneity of migratory phenomena, some of which have been known since antiquity, has stimulated various proposals for classification. Accidental and non-accidental migration, non-calculated and calculated migration, regular and irregular migration, long- and short-distance migration, horizontal or zonal and vertical or stratal migration, latitudinal and altitudinal migration, climatic, trophic and reproductive migration, daily, seasonal and megatemporal migration are only a few examples of the suggestions made by several authors in their endeavour to introduce order to the extraordinary variety and complexity of animal migration.

Migratory behaviour is widespread in desert and semi-desert environments. Desert soil animals must carry out vertical movements in response to temperature and moisture gradients and to escape from the widely fluctuating surface conditions (Wallwork 1982). These movements are often regulated by circadian rhythms. The vertical migration of subterranean animals, such as non-parasitic nematodes, pseudoscorpions, mites, collembolans, thysanurans, etc., has been studied in detail in woodland soil and leaf litter. Unfortunately, little attention has yet been paid to the study of vertical migration in desert soil animals. Daily zonal migration by species living on sandy beaches has also been little investigated. Only a few cases have been studied: e.g. isopods of the genus *Tylos* (Tongiorgi 1962, 1969), the mutillid wasp *Smicromyrme viduata* (Alicata et al. 1974) and the tenebrionid beetle *Pachyphaleria capensis* (Conti 1994; Fig. 44).

In contrast, the migration of locusts, first described in the Book of Exodus (1500 B.C.), has received considerable attention from several authors. Baker (1978) asserted that migration of the desert locust *Schistocerca gregaria* "has perhaps been studied in more detail than that of any other animal species". All locusts, living in tropical, subtropical arid and semi-arid areas, can effect mass migration. This is profoundly influenced by overcrowding. The South American *Schistocerca paranensis*, the Australian plague locust *Chortoicetes*

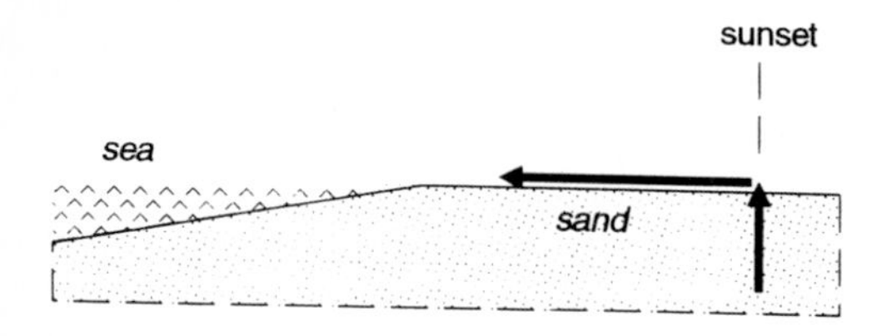

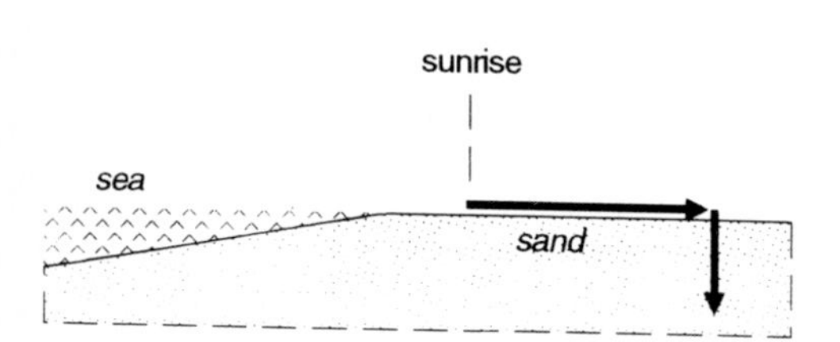

Fig. 44. Daily vertical and horizontal migration of *Pachyphaleria capensis*. (Adapted from Conti 1994)

terminifera, the African migratory locust *Locusta migratoria*, the Moroccan locust *Dociostaurus maroccanus*, the South African brown locust *Locustana pardalina* and the red locust *Nomadacris septemfasciata*, as well as *Schistocerca gregaria*, move in huge swarms searching for areas in which rain has fallen. Here, they feed on fresh vegetation and reproduce (Waloff and Rainey 1951; Rainey et al. 1957; Williams 1958; Gunn 1960; Michel 1969; Roffey 1972; Waloff 1972; Rainey 1976). Swarm cohesion is maintained by a highly gregarious tendency which originates in the premigratory phase. In many cases, locust migration is a kind of exodus without return (one-way migration). However, it sometimes takes place along more or less regular circuits which allow migrant locusts to re-enter their native area. For example, such migration has been shown by the desert locust (Baker 1978): the return migration seems to be a remigration, that is, a return migration by the offspring of the migrants on the outward journey. Moreover, in regard to the Rocky Mountain locust *Melanoplus spretus*, once extremely common in the arid northwestern territories of the United States and now practically extinct (Williams 1958), a regular seasonal to-and-fro migration was observed, southeast from July to September and northwest from May to June (Riley 1878, in Williams 1958). This true rectilinear migration does not agree with solely downwind movement. Indeed, Baker (1978) hypothesized the existence of preferred sun-compass directions among locusts.

Another well-known example of migration is shown periodically by hoofed mammals inhabiting open semi-desert areas. Some episodic and dramatic emigrations, such as those irregularly occurring in the past by springbok, have been described (see Cloudsley-Thompson 1978a). Zebras, wildebeest and Thomson's gazelles regularly migrate over the African savannahs, American bison over the North American prairies, saiga antelopes over the Eurasian steppes, reindeer and caribou respectively over the Eurasian and

North American tundras. These animals show clear, almost continuous nomadic habits, while, during the season in which trophic resources become depleted, they embark on impressive mass migrations. Herds consisting of tens to hundreds of thousands of individuals move along to-and-fro or subcircular routes, travelling long distances and maintaining characteristic formations (for example, wildebeests walk in single file, Talbot and Talbot 1963). The caribou and American bison may cover 650–800 and 650 km respectively on their annual journeys (Cloudsley-Thompson 1978a). Other large herbivorous mammals, such as the Canadian musk ox (Tener 1965), the eland (Grzimek and Grzimek 1960), the African elephant (Field 1971; Cloudsley-Thompson 1978a), the central Australian red kangaroo (Newsome 1965b), etc., have been described as being true migrants in past times. Some carnivores, such as lions, leopards, cheetahs, hyenas, wolves, hunting dogs, coyotes and jackals, following the movements of their potential prey, may present analogous partial migrations (see Schwartz and Schwartz 1959; Young 1964; Kruuk 1972; Schaller 1972; Baker 1978; Cloudsley-Thompson 1978a).

Many other animals that are habitual or occasional inhabitants or arid and semi-arid regions, such as millipedes, aphids, beetles, butterflies and moths, hymenopterans, toads, turtles and tortoises, birds, bats, and primates, may sometimes migrate. Among primates, it is right and proper to give human migration special attention. When they first invaded the African savannah (probably between the lower and upper Miocene), hominids were hunter-gatherers. After a long evolutionary history, the dichotomy between australopithecine and human forms took place among these savannah dwellers which began using tools 3 million years B.P. (Howell 1970). Soon after its emergence, the genus *Homo* began an impressive, and progressive emigration. It first spread throughout the Eurasian region and, about 32 000 B.P., *Homo sapiens* reached Australia (Jones 1973). Between 30 000 and 15 000 B.P. the species arrived in North America through the Beringia, a tundra landmass which at that time connected Siberia with Alaska. About 20 000 B.P., Man reached Central America and, between 20 000 and 10 000 B.P., northern South America (Cruxent 1968). The sub- sequent evolution of agriculture and animal husbandry progressively reduced the migratory habit. Modern hunter-gatherers, such as the Australian aborigines (Tindale 1953), the North Thailand Yumbri (Young 1962) and the Kalahari bushmen (Tobias 1964), are now confined to rather limited territories, but the nomadic habit still characterizes many human groups living in various arid and semi-arid areas of the world. Saharan Tuareg, Arabian Bedouin, Sudanese Fulani, East and South African Masai, Herero and Hottentots, are but a few examples of pastoral nomads in arid regions who move with their herds in a symbiotic relationship. Man provides his herds with good pasture, protection from predators, and assistance with parturition, whereas the herds supply Man with transport, milk, wool, and hides.

7 Exploitation of Food Resources

The exploitation of adequate food resources by animals is fundamental to life in every environment. From the trophic point of view, animals can be regarded as specialists or generalists. Specialists are monophagous or oligophagous species, i.e. they feed on one or only a few types of nourishment. Such animals may survive thanks to a marked ability to locate and utilize particular food. They are endowed with precise, genetically controlled systems to recognize nourishment. Moreover, they possess endogenous mechanisms which define the timing and duration of feeding. Their trophic patterns are usually invariable and, therefore, not susceptible to potential improvement by learning. In contrast, generalists are polyphagous, often omnivorous species (Fig. 45). They have a varied diet, based upon high nutritional versatility. Thus, they do not suffer from occasional or periodic shortage of food. Moreover, they are prone to investigate new trophic opportunities and, therefore, discover and learn to use new kinds of nourishment; they also explore and colonize new habitats.

Where food resources are scarce or scattered irregularly, as in a desert, the attainment of nutritional requirements imposes particular behavioural adaptations on animals. Opportunism is fundamental to take advantage of favourable circumstances whenever these occur. In arid environments, therefore, opportunistic trophism is widespread. Desert animals tend to be generalist feeders (MacArthur 1972; Louw and Seely 1982; Shmida et al. 1986). Even true carnivores, such as the Namib black-backed jackal (Stuart 1976), feed on vegetable matter when meat is scarce. Many herbivores feed on several species of plants; for example, the pronghorn antelope *Antilocapra americana* in the Chihuahuan Desert is able to exploit 168 species of forbs, 56 species of browse shrubs and 18 species of grasses (Buechner 1950). In addition, the diet may be supplemented with animal proteins and water (Sect. 7.1). For example, the granivorous North American desert sparrow *Amphispiza bilineata* augments its diet of dry seeds with insects, apparently for water balance (Serventy 1971). On the other hand, the diets of desert animals are often strictly dependent upon the need to regulate water balance (Reichman et al. 1979). Camels eat almost any vegetation that grows in desert and semi-desert regions; however, if very hungry, they will eat whatever is placed at their disposal: flesh, skin, bones and even fish (Walker 1968). In arid areas, specialists are also well represented. For example, adults of one Namib weevil

Fig. 45. The Palaearctic tenebrionid beetle *Pimelia grossa* is a typical omnivorous arid-adapted species

of the genus *Leptostethus* have a very limited diet, since they feed only on a perennial green grass, *Stipagrostis sabulicola* (Louw and Seely 1982). Giraffes are likewise oligophagous animals, feeding almost entirely on the leaves from acacia, mimosa and wild apricot (Walker 1968).

7.1 Herbivores

This category of animals is heterogeneous, and includes all animals that feed on plants, as well as detritivores (Sect. 7.3). There are several types of vegetable matter. These include (1) leaves, flowers and stems; (2) fruits; (3) roots; (4) seeds; (5) nectar and pollen; (6) sap; (7) wood; (8) bark; (9) detritus. Each of these resources requires the utilization of suitable morphological and behavioural patterns by the animals that feed on them. In dry environments, many animals are mainly herbivorous. Among these, particular attention is due to insects, rodents and ungulates, but many other invertebrates and vertebrates also deserve mention in this context.

7.1.1 *Insects*

Desert insects feed on all kinds of living plant tissues and/or seeds (Crawford 1981). Some of them eat ephemerals, others feed on perennial plants. According to Orians et al. (1977) and Cates (1981), the consumers of temporary desert vegetation are mostly generalists, while the consumers of woody perennial vegetation are almost evenly divided between specialists and generalists. Nevertheless, the lepidopteran larvae feeding on woody perennials are mainly specialists.

Orthopterans are mainly herbivorous. Their strong mandibles are adapted for cutting, tearing and triturating vegetable food. Most grasshoppers and locusts are generalist feeders. They feed on a wide range of plants and may represent a real danger to crops. For example, locusts, which show exceptional migratory ability (Sect. 6.2.3), are rightly considered to be the main insect pests of deserts. Even a moderate swarm can consume some 1000 tonnes of fresh vegetation daily (Uvarov 1966) ! Nevertheless, some grasshoppers are specialists: for example, the American species *Bootettix punctatus*, *Liguro-tettix coquilletti kunzei*, *Insara covilleae* and *Diapheromera covilleae* depend strictly on the creosote bush *Larrea divaricata* (Werner and Olsen 1973). Mole crickets and crickets are also polyphagous (Popov et al. 1984), but some desert crickets show a marked preference for certain plants. For example, the giant Namibian cricket *Brachytrupes membranaceus*, living in the dry bed of the Kuiseb river, feeds on the leaves and stems of *Acacia albida* and *A. giraffae biloba* (Fig. 46), very abundant trees along that peculiar linear oasis (Costa and Petralia 1990). Neanids and neoadults of this species collect large quantities of organic material from the ground near their burrows and transport it to their hole where they hoard it. They also utilize the leaves of *Euclea pseudebenus* and small pieces of *Stipagrostis ciliata*, *Datura innoxa*, and other herbaceous plants. Similarly, the Palaearctic sand cricket *Brachytrupes megacephalus*, living in coastal dune areas of Sicily, feeds essentially on the leaves and stems of *Lotus creticus* and *Thymelaea hirsuta*, which are stored inside its hole. However, it supplements its diet by eating inflorescences of *Plantago coronopus*, the fruit of *Pancratium maritimum*, bulbs of Liliaceae, and dried pods as well (Caltabiano et al. 1982a).

Different feeding habits are also found in desert Coleoptera. Many beetles are largely phytophagous. For example, Tenebrionidae and other families, such as Bruchidae and Anobiidae in various biotopes around Gobabeb in the Namib Desert are claimed to be phytophagous (Kuhnelt 1965, 1967). According to Endrody-Younga (1978), however, there are relatively few herbivorous tenebrionid beetles in the arid regions of southern Africa, while coprophagy appears to be the predominant feeding habit of these insects. The American species *Eupagoderes marmoratus* and *Eptcauta lauta* are listed by Werner and Olsen (1973) as feeding on *Larrea divaricata*. The leaves and stems of the North American mesquite *Prosopis juliflora* are the preferred food of the

Fig. 46. Acacia trees are very abundant and conspicuous along the fringes of dry riverbeds. They provide food for many desert animals

Bostrichidae (Riazance and Whitford 1974). Other specialist feeder beetles are the Chrysomelidae. The Palaearctic species *Timarcha pimelioides* (Fig. 47), of Sicilian coastal dunes (Costa et al. 1983b), is strictly oligophagous, only utilizing as food the leaves of *Plantago coronopus*, *Crucianella maritima*, *Launaea resedifolia* and *Rubia peregrina*. Adults show a marked preference for the first, and larvae for the second. The females also feed on dried vegetable debris during the mating and reproductive period.

Ants are very abundant in deserts. Many of them are herbivorous and eat leaves: the leaf-cutter ant *Acromyrmex versicolor* gathers the leaves of *Larrea divaricata* and both leaves and petals of various other plants in the Sonoran Desert as well as the blades of various species of grasses (Werner 1973). Other desert ants are granivorous. For example, some Saharan species of the genera *Messor*, *Monomorium* and *Pheidole* harvest and eat the seeds of grasses and other plants (Abushama 1984). The Mojave Desert species *Pheidole militicida* eats grass seeds almost exclusively (Creighton and Creighton 1960). The Chihuahuan Desert species *Pogonomyrmex desertorum* forages heavily on the seeds of *Tridens pulchellus* and *Bouteloua barbata*; other species of the genus *Pogonomyrmex* feed on the most abundant seed species present, but show a strong preference for *Eriogonium* spp. when available (Whitford 1973).

Fig. 47. The Palaearctic chrysomelid beetle *Timarcha pimelioides* is a typical specialist feeder

Termites are another conspicuous primary consumer in the desert. Their basic food is vegetable material (Wood 1977). The Saharan sand termite *Psammotermes hybostoma*, mentioned above as an exploiter of windblown debris (Sect. 3.1), can utilize all kinds of vegetation including the poisonous shrub *Calotropis procera*. It also damages wooden structures in semi-desert areas (Harris 1970; Abushama and Nour 1973; Hafez 1980; Nour 1980). The Sonoran Desert species *Heterotermes aureus*, *Gnathmitermes perplexus*, *Paraneotermes simplicicornis* and *Amitermes* spp. attack various types of wood, with a strong preference for the dead wood from *Acacia greggii*, *Cercidium floridum*, *Opuntia* spp. and *Prosopis juliflora* (Nutting et al. 1974; Haverty and Nutting 1975). According to Werger (1986), the most important phytophagous arthropods in the arid areas of southern Africa are termites of the genera *Trinervitermes* and *Hodotermes*. These feed especially on grasses, but also feed on various kinds of vegetable material. Some species even cultivate fungi in their nests (Ruelle 1978).

Many other herbivorous insects are found in deserts. Springtails, cockroaches, phasmatids, hemipterans, homopterans, lepidopterans and bees are primary consumers in desert environments.

7.1.2 Other Invertebrates

Very little is known about the feeding habits of desert snails, which are herbivorous or detritivorous, often feeding on living green plant material (Wallwork 1982). Some *Sphincterochila* spp. feed on surface mud, while certain species of the genus *Helicella*, which inhabit the Negev Desert, feed on the leaves of higher plants (Yom-Tov 1970; Schmidt-Nielsen et al. 1971).

Arachnids are typically carnivorous, however, some mites of the Mojave Desert are herbivorous (Wallwork 1972). In particular, the cryptostigmatid *Joshuella striata* and the astigmatid *Aphelacarus acarinus* and *Glycyphagus* sp. feed on fungal hyphae and spores, as well as decomposed leaf litter and plant detritus.

The social wood louse *Hemilepistus reaumuri* (see also Sect. 3.1) utilizes either green or dead pieces of *Hammada scoparia* and dry annual plants as food, as well as algae, fungi and lichens (Schachak et al. 1976).

Among centipedes, some Geophilomorpha may feed on vegetation, sometimes even damaging crops (Cloudsley-Thompson 1958). Millipedes are mostly phytophagous; the desert species *Orthoporus ornatus* feeds preferentially upon the superficial tissue of shrubs and dead plant material (Crawford 1974).

7.1.3 Rodents

The rodents inhabiting arid regions are mainly herbivorous. They consume practically every type of vegetable food. Moreover, they are often polyphagous, feeding on various species of plants. The majority are actually omnivorous, including quantities of insects in their diets.

Species belonging to the family Heteromyidae ingest dry seeds. Merriam's kangaroo rat *Dipodomys merriami* and the pocket mouse *Perognathus amplus*, which inhabit the Sonoran Desert (Reichman and Van der Graaff 1975), also eat greenery when available, and insects in somewhat different proportions in different years. The availability of food items certainly affects the diets of heteromyids, but clear alimentary preferences determine the exact amounts consumed by each species (Reichman 1975, 1978). Moreover, some heteromyids only rarely, if ever, drink free water (Hall and Linsdale 1929; Schmidt-Nielsen 1964; see also Sect. 2.5 and Chap. 4). Australian desert mice of the genera *Notomys* and *Leggadina* are strictly granivorous, yet are not dependent on the exploitation of free water (MacMillen and Lee 1967).

Among the Geomyidae (North American pocket gophers), the Pedetidae (South and East African springhaas), the Cricetidae (North American wood rats and Old World gerbils), the Chinchillidae (South American viscachas), and the Caviidae (South American maras and mountain cavies), there are desert and semi-desert species with primarily herbivorous habits. These ingest a variety of vegetation, including bulbs, roots, stems, grass, seeds, fruits and green leaves (with a prevalence of succulent plants) (see Vorhies and Taylor 1940; Jaeger 1948; Finley 1958; Petter 1961; MacMillen 1964; Schmidt-Nielsen 1964; Ryan 1968; Walker 1968; Dingman and Byers 1974; Reichman et al. 1979).

Many other arid-adapted rodents are omnivorous. They ingest both vegetable and animal food according to their preference at various times of year. For example, the Sonoran Desert cactus mouse *Peromyscus eremicus* usually feeds on insects, but it also ingests large quantities of seeds and green vegetation when available (Reichman 1978). The antelope ground squirrel *Spermophilus nelsoni* eats insects during summer and autumn, while it shows a decided preference for moist greenery in other periods of the year (Hawbreaker 1947). The Indian northern palm squirrel *Funambulus pennanti* is a vegetarian but, in summer, when little fruit is available, it feeds on locusts (Prakash and Kametkar 1969). The gerbils, too, are typically omnivorous with marked seasonal fluctuations in their alimentary preferences. For example, the Asian *Meriones hurrianae* eats locusts preferentially during the summer, leaves and flowers during autumn, seeds during winter, and stems and rhizomes during spring (Prakash 1962, 1969). This species, therefore, appears to follow a precise, regular dietetic programme according to the climate-dependent features of the environment. On the other hand, it is even able to change its activity rhythm, regulating it seasonally (Sect. 4.2). Other desert rodents are primarily carnivorous. They will be mentioned briefly in Section 7.2.2.

7.1.4 *Ungulates*

Ungulates are herbivorous, traditionally divided into grazers and browsers (e.g. Hoffmann and Stewart 1972). However, in arid regions, many species both graze and browse (Werger 1986). Moreover, they are distinguished on the basis of the site of digestion and fermentation: there are both forestomach and hindgut fermenters (Wilson 1989). The forestomach fermenters also include the ruminants, which belong to families of Artiodactyla (Camelidae, Cervidae, Giraffidae, Antilocapridae and Bovidae). According to Langer (1984), the ruminants can be divided into three major feeding classes: roughage, intermediate and concentrate selectors. Examples of arid-zone roughage feeders are species such as buffalo, cattle, and wildebeest (Fig. 48), which eat grass exclusively, and species such as hartebeest, gemsbok, oryx, and Thomson's gazelle, which eat grass (40% of the overall diet) as well as leaves, shoots and plant stems (60%). Arid-zone intermediate feeders are species such as sheep, eland, Grant's gazelle, impala, and springbok (Fig. 49), which feed mainly on leaves, blossoms, fruits and shoots (65%), but complete their diet with seeds, tubers and other reserve organs (35%).

Fig. 48. Wildebeest, *Connochates taurinus*, are roughage feeders, whose diet consists exclusively of grass

Fig. 49. Springboks are typical intermediate feeders

In contrast, desert concentrate selectors are species such as goat, camel, dik-dik, gerenuk, giraffe, kudu, and steenbok, which eat mainly seeds, shoots and blossoms (65%), supplementing their diet with leaves (35%). Several desert and semi-desert ungulates solve problems of food and water shortage by performing wide, almost continuous migratory displacements (Sect. 6.2.3). Many arid-adapted herbivorous mammals such as antelopes and other ungulates of the Kalahari Desert need to compensate for the deficiency of phosphorus and other minerals in their diet. For this reason, they exploit the 'salt licks' present in dry riverbeds and pans (Eloff 1963; Leistner 1967; Parris and Child 1973).

Among the Perissodactyla, African zebras (Fig. 50) and central Asian wild horses are well-known grass feeders. The white rhino is the only recent species of Rhinocerotidae which grazes rather than browses (Walker 1968).

Among the Artiodactyla, the collared peccary, which has been studied in southern Arizona (Eddy 1961), is a polyphagous herbivore: its diet includes fruits and the pads of *Opuntia* spp., the beans of *Prosopis* spp., fruits of *Cereus giganteus*, and grass in decreasing proportion; only rarely are insects eaten. The African wart hog feeds mainly on grass, roots, berries, the bark of young trees and occasionally also on carrion (Walker 1968). Camelids, which have a three-chambered, ruminant stomach, and specific ruminal Protozoa, exhibit the highest efficiency in comparison with the other ungulates in

Fig. 50. Zebras are typical grass feeders

digesting dry vegetable matter (Hintz et al. 1973; Wilson 1989). Some details of their diet were mentioned in the introductory part of this chapter. The mule deer of the Sonoran Desert eat succulent forbs and shrubs (Elder 1956); the Chihuahuan Desert subspecies, sometimes called Crook's mule deer, feeds preferentially on twigs and the leaves of bushes and low-growing trees, but it can also eat Cactaceae and Agavaceae (Sutton and Sutton 1966). *Acacia* trees are fundamental for the diet of several ungulates. The giraffe has no competitors due to its ability to browse acacia leaves at heights unattainable by other herbivores. The gerenuk shows a characteristic vertical browsing pose, standing upright on its hind legs to reach high acacia branches. Consequently, it is sometimes called the 'giraffe gazelle'. A similar pose for browsing on shrubs at considerable heights is exploited by the dibatag. Other herbivores, such as dik-diks, ingest large quantities of acacia leaves, but closer to the ground than do gerenuks and dibatags. In the Sudan, *Gazella dorcas*, which has a similar diet, does not appear to depend on drinking water (Carlisle and Ghobrial 1968). Other ungulates resort to different nutritional strategies. Many of them which normally feed upon grass can modify their diet during periods of drought when grasses are not available, e.g. the Indian gazelle eats plants such as *Capparis decidua*, *Crotalaria burhia*, *Aerva tomentosa* and *Calligonum polygonoides* which are avoided at other times (Gupta 1986). *Gazella granti* and *Oryx beisa* satisfy their need for water by feeding at night on hygroscopic plants (Taylor 1968). Even though they ingest large quantities of vegetable food, other ruminants, such as the gazelles of the Negev, need to drink free water (Shkolnik 1971). Many bovids show a bimodal rhythm of feeding. They lie in shelter during the heat of the day (Walker 1968). Kudu, goral and bighorn sheep (*Ovis canadensis*) graze and browse in the late evening and early morning; hartebeest, eland, and wildebeest forage during the morning and late afternoon or early evening. Other species, such as dik-dik and black buck, feed throughout the entire day. In contrast, pronghorn antelopes and impalas are active both at night and during the day, alternately grazing and/or browsing and resting.

7.1.5 Other Vertebrates

According to Miller (1932, 1955), the American desert tortoise *Gopherus agassizi* has strictly vegetarian habits. A few species of desert lizards, such as the Namibian dune lizard *Angolosaurus skoogi* (Hamilton and Coetzee 1969), are primarily herbivorous. Some of them, including Old World agamids of the genus *Uromastix*, are insectivorous as young and vegetarian when adult (Mertens 1960; Pope 1960). New World iguanids of the genera *Sauromalus* and *Dipsosaurus* are strictly herbivorous (Woodbury 1931; Shaw 1939, 1945; Norris 1953; Mayhew 1963a; Minnich and Shoemaker 1970; Pianka 1971; Nagy 1973; Hansen 1974); but, according to Aspland (1967), their juvenile forms feed on insects. Some other lizards, such as the ultrapsammophilous

Namibian lizard *Aporosaura anchietae* (Louw 1972) and the North American leopard lizard *Crotaphytus wislizenii* (Tanner and Krogh 1974) are highly opportunistic, feeding on both vegetable and animal matter (insects and other lizards).

Among granivorous birds in the arid areas of North America are Gambel's quail *Lophortyx gambelii* (Bartholomew and Cade 1963) and the white-winged dove *Zenaida asiatica*, the mourning dove *Zenaidura macroura marginella*, the house finch *Carpodacus mexicanus*, brown towhee *Pipilo fuscus*, and the black-throated sparrow *Amphispiza bilineata* (Tomoff 1974). The Australian budgerigar *Melopsittacus undulatus* (Cade and Dybas 1962) and zebra finch *Taenopygia castanotis* (Bartholomew and Cade 1963) are also described as seed-eaters. Many granivorous desert birds, however, must supplement their diet with greenery, succulent plants or free water (Lowe 1955; Bartholomew and Cade 1956, 1963; Bartholomew 1960; Guillion 1960; Smyth and Bartholomew 1966).

Most marsupials in arid-land ecosystems have herbivorous habits, but their diet is poorly known (Reichman et al. 1979). The central Australian red kangaroo *Megaleia rufa* specializes on green shrubs and grasses, with some preference for species of the genus *Eragrostis* (Chippendale 1962; Newsome 1965a).

Some primates and bats of desert areas are strictly vegetarian. For example, the entellus langur *Presbytis entellus* is a leaf-eating monkey (Reichman et al. 1979), as is the Indian pteropid *Pteropus giganteus*, a fruit-eating bat (Prakash 1959b). Lagomorphs are typical herbivorous animals. They usually feed on grasses and other herbaceous plants, but also eat bark when fresh vegetables are not available (Walker 1968). Desert hares (*Lepus* spp.), such as the well-known *Lepus californicus*, erroneously called the jackrabbit, differ from rabbits in that they do not dig burrows. Furthermore, their young are covered with fur, have their eyes open at birth, and are able to move and jump only a few minutes after birth. Rabbits, such as the desert cottontail *Sylvilagus auduboni*, have a diet based upon shrub browse and herbs (Chew and Chew 1970). Further analyses of the feeding habits of jackrabbits have shown that in the Mojave Desert they feed chiefly on annual leaf-succulent plants, such as *Salsola iberica*, and some perennial plants, such as *Larrea divaricata* and *Franseria dumosa* (Shoemaker et al. 1974). Turkowski and Reynolds (1974) discovered that in Arizona *Sylvilagus auduboni* is able to utilize 46 different plant species, especially grasses, and even various arthropods.

7.2 Carnivores

According to some scientists, carnivory is synonymous with predation, i.e. the process in which an animal spends time and effort to locate, mutilate or kill another living animal independently of its subsequent consumption

(Curio 1976). Carnivory is indeed a most comprehensive term. First, it also characterizes some plants. Carnivorous plants are present in humid areas of all continents: they are able to capture insects by sticky secretions or leaf traps and digest the prey by special enzymes. One of them (*Drosophyllum lusitanicum*) lives in arid areas of Portugal and Morocco (Polunin 1972). Moreover, the definition of carnivory and/or predation given above may also be applicable to parasitism. This, in my opinion, is a completely different process, and furthermore, it also is not limited to the animal kingdom. Finally, among carnivorous animals we must include those that feed on eggs, blood, or carrion.

According to the 'oversimplification' proposed by Crawford (1981), there are two main foraging tactics for carnivores: sit-and-wait and active hunting. Both predatory patterns are well represented in desert environments. In psammic areas, the sit-and-wait tactic is supported by the ease of detecting and locating seismic cues, especially when sandy soil is a medium suitable for conducting both surface and compressional waves (Brownell 1977). Ground-dwelling species, such as crabs (Horch 1971), scorpions (Brownell 1977; Brownell and Farley 1979), spiders (Den Otter 1974; Costa et al. 1993), ant lion larvae (Devetak 1985), anurans (Koyama et al. 1982; Lewis and Narins 1985), and reptiles (Hartline 1971; Hetherington 1989), may rely on vibratory cues for prey detection.

7.2.1 *Invertebrates*

Many desert invertebrates, particularly arthropods, are carnivorous (Fig. 51). Freckman and Mankau (1977) studied the distribution and trophic structure of the soil nematofauna of Rock Valley, and described many Dorylaimina as being predators or omnivores. Arachnids are mostly predatory. Many of them are chiefly insectivorous. Scorpions are broad generalists, and also frequently cannibalistic (Polis et al. 1981; see also Sect. 3.1). Many of them utilize sit-and-wait tactics (Stahnke 1966; Hadley and Williams 1968). Spiders, harvestmen, other scorpions, wood lice, myriapods, grasshoppers, crickets, mantids, cockroaches, earwigs, beetles, flies, butterflies, ants, and even small lizards and mice may be included in their diet (Cloudsley-Thompson 1958). The Californian Desert sand scorpion *Paruroctonus mesaensis* has been observed to accept 95 different prey species (Polis 1979). However, there are also species with clear food preferences, such as the Australian scorpionid *Isometroides keyserlingi* which specializes on burrowing spiders (Main 1956). Their ability to endure long periods without food enables scorpions to inhabit desert environments. For example, *Hadrurus* sp. may survive in the laboratory for 9 months without food or water (Stahnke 1945).

Various desert scorpions, such as the nocturnal North American species *Paruroctonus mesaensis*, are able to localize their prey by detecting substrate

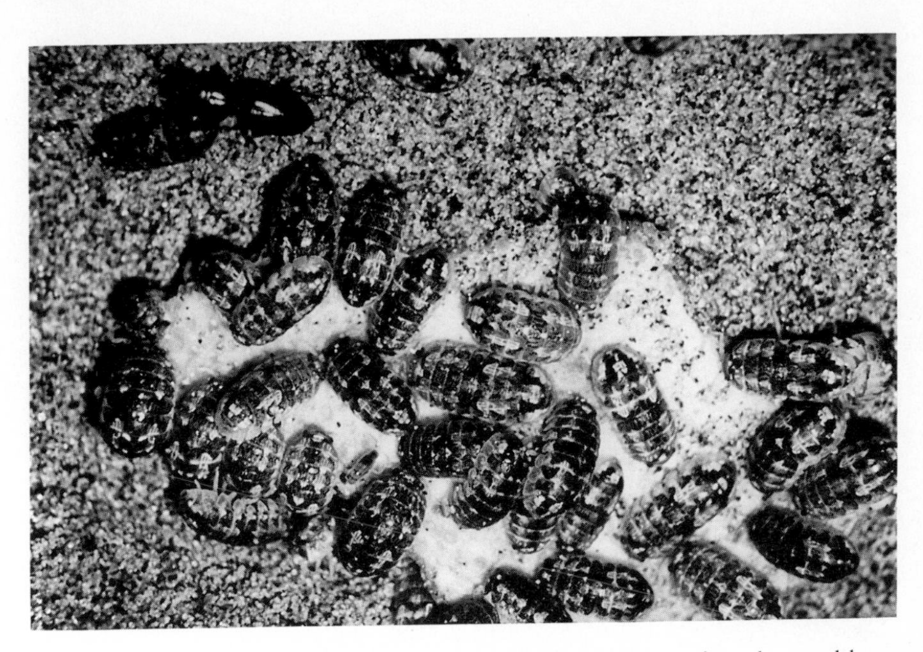

Fig. 51. Arthropods living on the seashore near the water's edge, such as the wood louse *Tylos latreillei*, can often exploit animal matter cast up by the waves. *Above* (to the *left*) Carabid beetles, *Scarites laevigatus*, which devour amphipods

vibrations (Brownell 1977; Brownell and Farley 1979). Solifugae are fast and voracious carnivores; they stuff themselves with food before the reproductive season. Insects (hard beetles included), spiders, scorpions, lizards, and sometimes even mice and small birds are killed and eaten (Cloudsley-Thompson 1958, 1977c). Feeding preferences have not often been noted, but some of the species that live in the North American deserts have been described as being dependent on termites as food (Muma 1966). The uropygids are also exclusively carnivorous. Their dietary habits are not yet well known, but captive specimens may feed on wood lice, cockroaches, grasshoppers, caterpillars, termites, centipedes, and even amphibians (Cloudsley-Thompson 1958; Wallwork 1982). Spiders are other notorious predators. Many insects, wood lice, myriapods and arachnids are potential prey. The predatory techniques of spiders inhabiting all ecosystems, including deserts, are numerous (Bristowe 1941). Some large Theraphosidae may even kill and eat vertebrates such as amphibians, lizards, young snakes, birds and small mammals (Millot 1943; Cloudsley-Thompson 1958). The large heteropodid *Carparachne* sp., which lives in sandy areas of the Namib Desert, seems to prey mainly on the web-footed dune gecko *Palmatogecko rangei* (Lawrence 1959). Energy-based territoriality, analyzed in populations of the North American desert funnel-web spider *Agelenopsis aperta* (Riechert 1978), is probably widespread among desert spiders, according to Cloudsley-Thompson (1991). Moreover, the size of the territory appears to be inversely proportional to prey abundance. Some aganippine trapdoor spiders are able to expand their area of capture by attaching twigs to the rims of their holes (Main 1957). Among the non-parasitic desert mites, there are predatory species. For example, adult giant velvet mites (*Dinothrombium* spp.), present in both American and African arid zones, are carnivorous, probably feeding on termites and other insects (Cloudsley-Thompson 1962, 1975), while their larval forms are parasitic on grasshoppers (Tevis and Newell 1962). Prostigmatid mites living in the Chihuahuan Desert seem to be predators of soil nematodes (Santos et al. 1978), while chilopods are also mainly carnivorous. They feed on worms, wood lice, insects, toads, geckoes, and some even kill small birds and mice (Cloudsley-Thompson 1958). Only the largest species are able to prey on vertebrates, however, and these are inhabitants of the wet tropics.

Among insects, several species belonging to various groups have carnivorous habits. Mantids, ant lions, carabids, cicindelids, and wasps are typical examples. Scarabaeids, trogids and geotrupids are well known as dung feeders, dipterans and some hemipterans as blood suckers, and tenebrionids and other beetles as carrion feeders (Sect. 3.2.11).

7.2.2 Vertebrates

All desert vertebrate groups include carnivorous species among their members. Adult amphibians are primarily generalist insectivores and this is true of arid-adapted species, such as the Australian desert frogs (Calaby 1960). The South African species *Pyxicephalus adspersus* eats termites and ants (Fitzsimons 1935) as well as other frogs, small snakes, ducklings and mice (Rose 1962; Mayhew 1968). The tadpoles of certain species, such as *Scaphiopus bombifrons*, *S. holbrooki*, *S. couchi*, *S. hammondi*, and *Pyxicephalus adspersus*, are cannibalistic (Bragg 1946, 1957, 1962, 1964; Mayhew 1965; Orton 1954; Rose 1962).

The North American desert tortoise *Gopherus agassizi*, usually described as strictly vegetarian (Sect. 7.1.5), nevertheless eats snails in captivity (Nichols 1953). Most lizards are insectivorous. Some of them, such as the Australian agamid *Amphibolurus adelaidensis* (Tyler 1960), are generalist feeders, while others specialize on termites or ants. For example, the South African lacertids *Eremias namaquensis* and *Ichnotropis squamulosa*, and the gekkonid *Palmatogecko rangei* (Fitzsimons 1935), as well as the North American teiid *Cnemidophorus hyperythrus beldingi* (Bostic 1966) eat termites almost exclusively. Similarly, North American iguanids of the genus *Phrynosoma* (Bryant 1911; Little and Keller 1937; Norris 1949; Reeve 1952), the Australian agamid *Moloch horridus* (Davey 1923; Hosking 1923; Barrett 1928; White 1948) and the South African lacertid *Eremias lineo-ocellatus* (Fitzsimons 1935) are chiefly ant eaters. Some lizards may regularly feed on vertebrates. For example, the diet of the North American gila monster *Heloderma suspectum* includes both bird and reptile eggs as well as young rodents (Shaw 1948; Hensley 1949; Bogert and Martin del Campo 1956). Cannibalistic habits have been described for several species, such as the gekkonid *Hemidactylus flaviviridis* (Rao 1924; Mahendra 1936), the iguanids *Crotaphytus wislizenii* (Ruthven 1907; Montanucci 1965), *Phrynosoma douglassi* (Dodge 1938), *Sceloporus orcutti* (Mayhew 1963a), *S. occidentalis* (Johnson 1965), *Uma notata* (Shaw 1950), the varanid *Varanus griseus* (Flower 1933) and the xantusid *Xantusia vigilis* (Heimlich and Heimlich 1947).

Snakes are typically carnivorous. Insects and other small invertebrates form the basic diet of desert colubrids such as the North American *Chionactis occipitalis* (Klauber 1951) and the South African *Prosymna sundevalli* (Fitzsimons 1962). Some snakes specialize on termites: examples are the North American western blind snake *Leptotyphlops humilis* (Klauber 1940) and the South African worm-snake *Typhlops schlegelii* (Fitzsimons 1962). Other species feed mostly on vertebrates. For example, the North American sidewinder rattlesnake *Crotalus cerastes* (Funk 1965) and the Namib Peringuey's adder *Bitis peringueyi* (Fitzsimons 1962) show clear preferences for lizards. According to Leopold (1961), the diet of some Australian pythons is based upon wallabies and kangaroos. The North American desert gopher snake (*Pituophis catenifer*) often preys upon jackrabbits, especially young ones (Sutton and Suton 1966) and rodents (Klauber 1947). Other arid-adapted snakes, such as the South African puff adder *Bitis arietans* (Fitzsimons 1935), the South African colubrid *Boaedon fuliginosus* (Fitzsimons 1962), the speckled rattler *Crotalus mitchelli* (Klauber 1936) and the North American long-nosed snake *Rhinocheilus lecontei* (Klauber 1941), prefer rodents. Some snakes feed on eggs. The non-poisonous South African egg-eating snake *Dasypeltis scabra* feeds exclusively on birds' eggs (Gans 1952; Fitzsimons 1962; Rose 1962), while the North American leaf-nosed snake *Phyllorhynchus decurtatus* (Brattstrom 1953; Miller and Stebbins 1964) and the South African elapid *Elapsoidea sundevalli* (Fitzsimons 1962) show a preference for lizard eggs. Many snakes are ophiophagous. The South African elapid *Naja anchietae* preys on colubrids of the genus *Psammophis* (Fitzsimons 1938) and the viperid *Bitis arietans* (Rose 1962). The North American colubrid *Masticophis taeniatus* eats the western diamondback rattlesnake *Crotalus atrox* (Stebbins 1954); and so on. The North American southwestern ringneck snake *Diadophis punctatus* preys on both earthworms and reptiles (Gehlbach 1974). Various species, belonging to the families Colubridae, Elapidae and Viperidae, are also cannibalistic (Fitzsimons 1935, 1962; Funk 1965; Miller and Stebbins 1964).

Three prevailing types of feeding habits are shown by desert birds. In decreasing order of numbers, these are insectivorous, granivorous and non-insect-eating carnivorous (Cloudsley-Thompson 1975). In a study of the ornithofauna of California, Miller (1951) established that most of the birds associated with deserts feed on insects. This diet guarantees the moisture necessary for the survival of arid-adapted species. Tomoff (1974), when analyzing avian species diversity in desert scrub communities of the Sonoran Desert, described seven foraging categories, five of which refer to insect-eating. According to this author, gilded flickers *Colaptes chrysoides*, Bendire's thrashers *Toxostoma bendirei*, curve-billed thrashers *Toxostoma curvirostre*, and crissal thrashers *Toxostoma dorsale* are examples of ground insect feeders; verdins (*Auriparus flaviceps*) and black-tailed gnatcatchers *Polioptila melanura* are foliage insect feeders; gila woodpeckers *Centurus uropygialis* are typical bark insect feeders; cactus wrens *Campylorhynchus brunneicapillus* and rufous-winged sparrows *Aimophila carpalis* feed on both ground and foliage insects; ash-throated flycatchers *Myiarchus cinerascens* prey on aerial insects.

The verdin *Auriparus flaviceps*, studied in Arizona by Taylor (1971), includes spiders as well as insects in its diet. African nightjars (*Caprimulgus* spp.) and some swallows and swifts appear to be specialized termite-eaters (Serventy 1971). The Sonoran elf owl *Micrathene whitneyi*, which feeds almost entirely on arachnids, centipedes and insects, is also able to seize lizards and snakes as an adult (Ligon 1968). The roadrunner normally preys on arthropods and diurnal lizards (e.g. *Cnemidophorus* spp.), however, the oldest individuals concentrate on lizards (Bryant 1916; Ohmart 1973).

Vertebrate-eating desert birds utilize a great variety of prey. For example, the Chihuahuan owl *Bubo virginianus* has been described as a predator of eight different mammal species, belonging to Rodentia (*Mus musculus*, *Peromyscus eremicus*, *Sigmodon hispidus*, *Reithrodontomys* sp., *Dipodomys merriami*, *Peromyscus* sp.), Insectivora (*Notiosorex crawfordi*) and Chiroptera (*Macrotus californicus*) (Bradshaw and Howard 1960; Reichman et al. 1979). Other desert birds of prey, such as *Falco peregrinus babylonicus* (Bartholomew and Cade 1963), often eat bats. In desert water holes, *Falco biarmicus* lies in ambush waiting for columbiformes when they alight on the water (Cade 1965).

Only a few marsupials, such as the Patagonian opossum *Lestodelphis halli* (Sect. 3.2.16) and the Australian mulgara *Dasycercus cristicauda*, are primarily carnivorous. The diet of the mulgara includes insects, small reptiles and rodents (Schmidt-Nielsen 1964).

The food of desert Insectivora consists mainly of insects. The diet of the North American desert shrew *Notiosorex crawfordi* is based upon insect larvae, orthopterans, dipterans and lepidopterans, and may include centipedes (Fisher 1941; Hoffmeister and Goodpaster 1962). In addition to a variety of insects (ortopterans, beetles, termites, neuropterans), the Indian hedgehog *Hemiechinus auritus* may also ingest toads, lizards, mammalian carrion and eggs (Krishna and Prakash 1956; Prakash 1956, 1959a, b). The Namib Desert golden mole *Eremitalpa granti namibensis* consumes a wide variety of invertebrates (mostly termites and coleopteran larvae and, in lesser proportion, other insect larvae, spiders, lepismatids, ants and beetles) and sometimes even Scincidae (Fielden et al. 1990).

The East African hamadryas baboon *Papio hamadryas* is omnivorous. In addition to a variety of plants, it also ingests large quantities of locusts and even shreds of dik-dik carrion (Kummer 1968).

Microchiropteran bats are often exclusively insectivorous. The American species *Eumops perotis*, *Tadarida brasiliensis* and *Antrozous pallidus* are generalist feeders, eating several species belonging to various insect orders, such as Hymenoptera, Lepidoptera, Hemiptera, Orthoptera, Homoptera, Neuroptera and Coleoptera (Ross 1961). The Indian false vampire *Megaderma lyra* may also prey on other bats, lizards and birds, and, in captivity, on sparrows, gerbils, mice and rats (Prakash 1959c).

The armadillos living in North American arid areas are carnivorous, feeding on insects, amphibians, lizards, snakes and eggs (Walker 1968). South American edentates, such as *Zaedyus pichyi*, are omnivorous, ingesting vegetables and fruits, insects, snakes, lizards, and mammals in decreasing proportion (Reichman et al. 1979).

Several desert rodents are at least partially insectivorous (Sect. 7.1.3). Some carnivorous mice, such as the North American grasshopper mice *Onychomys* spp., supplement their insectivorous diet with birds and mammals (Bailey and Sperry 1929; Schmidt-Nielsen and Haines 1964; Flake 1973). The Negev spiny mice (*Acomys* spp.) are specialized snail-feeders (Shkolnik and Borut 1969; Shkolnik 1971).

The fresh meat ingested by Carnivora enables these animals to be independent of free water and therefore preadapted to desert life (Bartholomew and Dawson 1968). Moreover, this social organization and aptitude for communal hunting eases survival under hard conditions. Coyotes (*Canis latrans*) are generalist feeders, inclined to omnivory. Their diet may be strictly carnivorous (Sperry 1941) or include up to 80% vegetable material (Murie 1951). It consists essentially of vertebrates (such as jackrabbits and rabbits, rodents, and collared peccaries), insects (such as grasshoppers and beetles) and plants (such as *Acacia*, *Opuntia* and *Prosopis* spp.) (Murie 1951; Walker 1968). Other mammals, such as sheep and goats, are occasionally killed. Hunting is performed singly or in relays. Individual coyotes sometimes form a particular, not yet well understood, interspecific hunting partnership with individual badgers (*Taxidea taxus*) (Walker 1968). Jackals also have a mixed diet, feeding on small, sick or injured medium-sized vertebrates, carrion, invertebrates and vegetables (Walker 1968). For example, the Indian jackal *Canis aureus aureus* feeds readily on bovid carcasses or preys on beetles, scorpions, gerbils, mongooses, ailing sheep and gazelles (Prakash 1969). Moreover, it may ingest a quantity of vegetable matter, crops included (Prakash 1959b). Desert foxes consume both animal and vegetable food. The North American kit fox *Vulpes macrotis* preys on rabbits, kangaroo and wood rats, pocket mice, various cricetids, small birds, lizards, insects and scorpions (Egoscue 1962; Sutton and Sutton 1966). The red fox *Vulpes vulpes*, studied in the Indian deserts, feeds on gerbils, insects, scorpions, berries and the seeds of Cucurbitaceae and Rhamnaceae (Blanford 1888–1891; Reichman et al. 1979). The Bengal fox

Vulpes bengalensis eats scorpions, insects, ground birds, lizards, gerbils and seeds of *Citrullus vulgaris* (Pocock 1941; Prakash 1969). The Saharan fennec also exploits a variety of foods such as insects, lizards, birds, birds' eggs, small mammals and plant material (including dates) (Schmidt-Nielsen 1964; Walker 1968). Arid-region mustelids are also carnivorous, with a propensity for omnivory. For example, according to Martin et al. (1951), the North American badger *Taxidea taxus* almost certainly ingests only animal matter, e.g. ground squirrels (*Citellus* spp.), smooth-toothed pocket gophers (*Thomomys* spp.), rabbits, lizards and insects. In contrast, the American skunks *Mephitis mephitis*, *Mephitis macroura* and *Conepatus mesoleucus*, consume varying amounts of vegetation in addition to insects, spiders, frogs, lizards, mice and eggs of birds. On the other hand, viverrids, hyaenids and felids are strictly carnivorous and only rarely supplement their diet with vegetation. For example, only 4.5% of the diet of the North American bobcat *Lynx rufus* is composed of vegetables (Young 1958). The Indian mongoose *Herpestes edwardsi ferrugineus* feeds on rodents, partridges, varanids, insects and scorpions (Prakash 1969), while the small Indian mongoose *Herpestes auropunctatus pallipes* consumes insects and scorpions almost exclusively (Reichman et al. 1979). Except for the aardwolf, which feeds mainly on termites and insect larvae, the hyaenids are primarily scavengers or predators of vertebrates. The spotted hyaena *Crocuta crocuta* is even able to kill old lions and half-grown rhinoceroses; at the same time, it succeeds in damping its prodigious appetite by ingesting large numbers of locusts (Walker 1968). The Indian desert cat *Felis libyca ornata* consumes both gerbils and insects (Prakash 1969). The puma *Felis concolor* eats rodents, rabbits, deer and carrion (Young 1946). The largest felids, availing themselves of their great strength, high speed and, often, suitable social organization, prey on almost any vertebrate that they come across.

7.3 Detritivores

According to Odum (1971), the term 'detritus' refers to organic matter produced by the decomposition of dead organisms. From an ecological point of view, it is a fundamental link between the living and the inorganic world (Odum and De la Cruz 1963). Many invertebrates, belonging to several different taxa, consume organic debris, microflora and fungi. They have detritivorous habits and play important roles in the energy and nutrient flow of all ecosystems, interacting with bacteria and fungi in the breakdown and decomposition of organic matter. In terrestrial environments, these animals are usually associated with the soil. Some general features of soil-bound desert detritivores and specialized exploiters of windblown debris were mentioned in Section 3.1. Further information was furnished in Section 3.2.

A very interesting review of life history patterns and trophic roles of invertebrate desert detritivores was made by Crawford (1979). He broadly divided these animals into three main groups: (1) short lives (annual or sub-annual species); (2) long lives; (3) intermediate situations. The life histories of these categories of detritivores differ not only in longevity, but also in patterns of responsiveness to external conditions and, above all, to rainfall. The distribution, abundance and activity of these animals are also affected by other factors, including temperature, the location of nutrients, soil features and topography. Short-lived species, which are also able to respond rapidly to changing environmental conditions, include nematodes, mites, collembolans, thysanurans, cockroaches, and some orthopterans, coleopterans (larvae and/or adults) and dipterans (mostly larvae). Nematodes appear to be particularly abundant and significant in the surface layers of both polar (Wallwork 1976; Chernov et al. 1977) and warm deserts (Freckman and Mankau 1977). Some beetles and flies are specialized for dung and/or carrion feeding. Short-lived species tend to consume fresh matter, taking the maximum advantage of existing moisture (Matthews 1976). Long-lived species, with variable responses to environmental change, include snails, millipedes, and the larvae of insects such as root-feeding lepidopterans and wood-boring beetles. The Negev snail *Sphincterochila boissieri* (Schmidt-Nielsen et al. 1971; Shachak et al. 1976) and the North American millipede *Orthoporus ornatus* (Crawford 1974, 1976; Wooten and Crawford 1974) are particularly well-studied desert detritivores. The intermediate group includes mainly isopods and social insects such as termites and litter-feeding ants. These insects appear to have a fundamental, direct influence on nutrient and energy flow in warm deserts (Watson et al. 1973; Nutting et al. 1975; Matthews 1976; Nutting and Haverty 1976; Mares and Rosenzweig 1978; Whitford 1978a).

8 Reproductive Regulation

Breeding and reproduction are strongly influenced by the desert environment. Thermal extremes, irregularity of rainfall and the availability of food are the main extrinsic factors that make the reproductive success of arid-adapted species difficult (Riechert 1979). After MacArthur and Wilson (1967) and Pianka (1970), the reproductive adaptive patterns of animals are usually subdivided into two alternative survival strategies (or types of selection), according both to environmental and biological parameters: '*r* strategy' and '*K* strategy'. *r* strategists are opportunistic species, adapted to exploit the ephemeral resources available in a harsh environment with unsettled and unpredictable climatic conditions. According to Horn (1978), these species are relatively small and short-lived; they may have a strong emigratory tendency and be rapid colonizers of new habitats. They are usually semelparous and produce many offspring which develop rapidly and do not require high parental investment. *K* strategists, in contrast, are large and long-lived species characterized by steady and often crowded populations. They are usually iteroparous and produce few offspring at a time. Moreover, parental care among them is pronounced and prolonged.

As we consider environmental unpredictability to be a basic factor, we would expect to find a strong prevalence of *r* strategists over *K* strategists among desert species. This is particularly true of animals that are almost continuously surface-active or dwell in temporary waters; but it is not true of species which inhabit a more or less stable subterranean habitat, because climatic parameters there present only minor diurnal and seasonal fluctuations, as in non-desert ecosystems. In addition, no single biological factor determines absolutely a particular survival strategy. For example, even though a relatively large body size might appear to be advantageous to desert-dwelling animals, since it minimizes water loss to the environment, numerous species, especially those belonging to the soil fauna such as snails, wood lice, microarthropods, and ants, do not conform to this trend (Wallwork 1982).

Scorpions, which are considered to be one of the animal groups best adapted to desert environments, are large, long-lived and iteroparous (Polis and Farley 1980). Nevertheless, some of them, such as the North American *Paruroctonus mesaensis* (Polis 1979) and *Centruroides sculpturatus* (Williams 1969), produce medium-sized clutches, while other species, including the North African *Androctonus australis*, *Leiurus quinquestriatus* and *Buthus occitanus*, produce large clutches (Polis and Farley 1979b). The long-lived Australian scorpion *Urodacus yaschenkoi* is semelparous and produces very small clutches (Shorthouse 1971). Among desert rodents, there is also a variety of survival strategies. Heteromyids and sciurids are usually *K* strategists, murids are *r* strategists, and cricetids occupy an intermediate position (French et al. 1975; Wallwork 1982). Nevertheless, various exceptions to these generalizations are known. For example, the South African big-eared desert mouse *Malacothryx typica* (Muridae) appears to be somewhere in the middle of the *r–K* range (Knight and Skinner 1981). It is not wise, therefore, to generalize survival strategies within a particular taxonomic group: on the other hand, it must be admitted that *r* and *K* strategies are two extreme selection patterns in an almost continuous spectrum of adaptive possibilities.

8.1 Timing of Reproduction

Animals usually breed during more or less restricted periods of the year. Choice of season depends closely on life cycle and the habitat of each species, and must assure the most favourable environmental conditions for survival of the offspring. An important factor of reproduction is parental care. This is particularly true in species without overlapping generations. In certain situations, the survival of the parents may be essential for that of their offspring (Louw 1993). Moreover, complex social organization may exceed the importance of seasonal breeding (Millar 1972).

Seasonal timing of the reproduction of desert animals is generally synchronized by external factors, such as rainfall (and subsequent food availability), photoperiod and temperature (Cloudsley-Thompson 1991). Breeding often occurs during the rainy season in arid environments, sometimes just before the rains fall that guarantee food for the young. Among birds, examples include the woodpeckers and parrots that live in the Sonoran Desert. These nest just prior to the summer rainy season and, consequently, their brood can profit from emerging insects (Short 1974). Among arachnids, the New Mexican spider *Agelenopsis aperta*, which inhabits desert-grassland areas, lays its eggs just before the rains (Riechert 1974). While many adults die as a result of flooding, the embryos are protected by the egg sacs and survive. Later, the young enjoy an abundance of prey (Riechert and Tracy 1975). Different periods of reproduction may occur in the same species in desert areas with differing climatic and topographic features: the species shows great behavioural plasticity (Riechert 1979). Certain desert-dwelling animals, such as insects, spiders, birds, mammals, and most amphibians, breed only after the rainy season has begun. This happens in the African red locust *Nomadacris septemfasciata* (Woodrow 1965), in the Chihuahuan leaf-cutter ant *Acromyrmex versicolor* (Werner and Murray 1972), and in the Namibian larks (Willoughby 1971).

In desert snails (Cloudsley-Thompson and Chadwick 1964; Schmidt-Nielsen et al. 1971; Yom-Tov 1972), amphibians (Buxton 1923), birds (Moreau 1950; Keast and Marshall 1954; Serventy and Marshall 1957; Keast 1959; Immelmann 1963) and marsupials (Newsome 1966; Brown 1974), breeding may occur at any time of the year, provided rain falls and the temperature is moderate. According to Immelmann (1963), the very sight of rain triggers courtship in many Australian birds.

Various arid-adapted species breed throughout most of the year, but show a significant increase in reproductive activity at the time of rainfall. Such breeding peaks have been described in grasshoppers (Chapman 1962), Lepidoptera (Hsiao and Kirkland 1973), amphibians (Main et al. 1959; Bragg 1965), marsupials (Frith and Sharman 1964; Sadlier 1965; Ewer 1968a; Brown 1974) and rodents (Prakash et al. 1971). Species that live in arid regions with two distinct periods of rainfall may correspondingly have two breeding periods. This occurs in the Chihuahuan tiger salamander *Ambystoma tigrinum* (Webb 1969) and in Sonoran birds, such as sparrows, thrashers and the roadrunner (Ohmart 1969, 1973; Short 1974).

It is occasionally possible to note a delayed reproductive response to the rainy season. For example, the Mojave sand scorpion *Paruroctonus mesaensis* (Polis and Farley 1979b) mates in spring, while the rain falls in winter. Nevertheless, the female retains and nourishes the embryos within her uterus for a long time, and birth does not take place until the following summer. Another example can be seen in the same desert. The oribatid mite *Haplochthonius variabilis* (Wallwork 1972) produces a new generation some 3 to 4 months after the winter rains have fallen and before the higher spring temperatures arrive.

Photoperiod, which changes with absolute regularity from year to year, is probably the most reliable environmental cue for controlling many rhythmic biological activities (Kendeigh 1961). Reproduction may therefore be timed by changing the length of daylight. For example, among various desert birds and small mammals, the extension of the photoperiod in spring, which indirectly is often a sign of probable rainfall in the near future, induces an increase in gonad size. These animals maintain subthreshold sexual activity until the rain falls, stimulating the growth of plant seedlings (Louw 1993). In other species, including certain Indian rodents, spring reproduction, triggered by day length, appears to be independent of moisture or nutritional factors (Prakash 1971). Photoperiod is also an important factor for the seasonal timing of breeding in arthropods and reptiles (Cloudsley-Thompson 1991).

The reproductive activity of the Palaearctic cricket *Brachytrupes megacephalus* is synchronized with the onset of spring in Sicilian coastal dunes (Caltabiano et al. 1982a). Various external factors may concur in determining the timing of reproduction of this species there, but the main factor appears to be the increasing minimum temperature values after the rigors of winter. In Sicily, spring coincides with the onset of the arid period as well as lengthening photoperiod. Furthermore, it has been ascertained that the presence of wind inhibits the stridulatory activity of the males. Reproductive activity of the Namibian giant cricket *Brachytrupes membranaceus* begins towards the end of January (Costa et al. 1987). In this species, the importance of both temperature and wind in regulating the singing of the males has also been demonstrated.

Other environmental cues may control reproduction in arid-adapted animals. For example, the lunar cycle appears to affect the sexual activity of the impala (Skinner and Van Jaarsveld 1987). The restricted breeding period of this antelope is triggered by the rutting calls uttered by males during the May full moon.

8.2 Partner Location

The problem of partner location has not been extensively analyzed in any ecosystem, particularly in dry environments. Nevertheless, meeting of the sexes is essential for propagation. It depends mainly on the spatial and temporal distribution of the sexual partners in their habitat. Partner location is achieved through specific communication signals. Releaser pheromones and acoustic messages are widely employed to attract potential partners from large distances (Sect. 6.2.2).

Attractant pheromones are generally emitted by females and induce chemotactic responses in males. Among terrestrial animals, they occur in both invertebrates (e.g. arachnids and insects) and vertebrates (e.g. reptiles and mammals). Males of short-sighted, hunting spider species may utilize their chemotactic sense to locate the females (Cloudsley-Thompson 1958). Sex attractants of insects may be released into the air or deposited on the ground (Butler 1970). Aerial trails are used by flying insects (Wright 1958; Bossert and Wilson 1963), terrestrial trails characteristically by ants and termites (Wilson 1971; Moore 1974). In the North American western banded gecko *Coleonyx variegatus*, chemical stimuli seem to be involved in the meeting and recognition of the sexes (Greenberg 1943). Sexual dimorphism is slight among snakes (Davis 1936) and odour has been hypothesized as an important factor in the early phases of reproductive behaviour (Davis 1936; Stebbins 1954; Klauber 1956; Bellairs 1959; Miller and Stebbins 1964). On the other hand, it appears likely that both the nose and Jacobson's organ enable male vipers and other snakes to follow odours released by the cloacal glands of the females (Bellairs 1969). A recent study of the courtship behaviour of the Mexican garter snake *Thamnophis melanogaster* (Crews and Garstka 1982) has shown that the vitellogenin produced by well-fed females attracts males and triggers male courtship.

Among those mammals that are endowed with a Jacobson's organ, attractant pheromones may be important cues for assisting both the meeting of the sexes and checking the partners' sexual states (Kevles 1986). In many placentals, including Perissodactyla, Artiodactyla and Carnivora, males reply with a particular facial expression, named 'flehmen' by German ethologists, to odours produced (often, through the urine) by an oestrous female: "the head is raised, the lips turned back and the nose wrinkled and breathing is stopped for a moment" (Ewer 1968b). Flehmen appears to indicate the olfactory analysis of the female's sexual readiness. In the southern African black rhinoceros, *Diceros bicornis*, an odorous substance, present in the cow's urine, attracts the bull to the mating site (Goddard 1966; Schenkel and Schenkel-Hulliger 1969). Among rodents, both releasing and priming reproductive pheromones have been well studied (Sect. 8.3). For example, messages originating from females which attract males, and male signals which attract females, have been described among mice (Bronson 1974). Olfactory attraction of males to a receptive female has also been reported in deer mice (*Peromyscus* spp.) (Doty 1972). Other examples will be discussed in Section 8.3.

Stridulation is widespread in the animal kingdom. It has been recorded in many non-insect arthropods, such as centipedes (Cloudsley-Thompson 1958), scorpions (Alexander 1958, 1960; Dumortier 1964; Constantinou and Cloudsley-Thompson 1983), solifuges (Dufour 1862; Hansen 1893; Dumortier 1964; Cloudsley-Thompson and Constantinou 1983), spiders (Bristowe 1958; Cloudsley-Thompson 1958) and isopods (Caruso and Costa 1976). Nevertheless, in most cases, sound production is not linked with the faculty of hearing. It should therefore not be interpreted as an acoustic call: it is probably only a form of advertisement or a warning to enemies (Cloudsley-Thompson 1958). Sometimes, as in scorpions, a certain degree of hearing may be present, but it probably has the anti-predatory function of detecting ground vibrations (Cloudsley-Thompson 1955). Acoustic calls are usually made by males. Among many species of insects belonging to different taxa, the meeting of mates is based on a system of male calling and female phonotactic reply.

The importance of male courtship song, sometimes linked with territorial habits, is apparent among orthopterans, both grasshoppers and crickets (Fig. 52). The mature male of the giant Namibian sand cricket *Brachytrupes membranaceus* constructs an 'open sky chamber' at the entrance to its hole. This amplifies the song to which females are attracted (Costa et al. 1987). Similar behaviour is performed by the Sicilian sand cricket *Brachytrupes megacephalus* (Caltabiano et al. 1982) and by the French Polynesian short-tailed cricket *Anurogryllus muticus* (Walker and Whitesell 1982). In all these cases, the females are guided to their mate by stridulation. Among sand acridids, male singing and female phonotaxis are also employed in the meeting of the sexes (Costa and Messina 1974). In the gomphocerine grasshopper *Bootettix argentatus*, which lives and feeds exclusively on leaves of the creosote bush *Larrea divaricata* in the Sonoran Desert, males attract females with acoustic signals emitted both day and night (Otte 1970). Groups of the Chihuahuan Desert gomphocerine tarbush grasshopper *Goniatron* (=*Ligurotettix*) *planum*, which lives almost exclusively on southern

Fig. 52. A male of the Palaearctic cricket *Brachytrupes megacephalus* stridulates when it is sexually mature (the spermatophore is visible)

blackbrush *Flourensia cernua*, are comprised of both singing and silent males (Otte and Joern 1975). On each bush there is only one singing male and several silent males. The latter are able to intercept some of the females that are attracted to the singing male, without having to expend energy or to reveal their presence. A series of studies on another desert grasshopper (*Ligurotettix coquilletti*), living in shrubby areas of the Sonoran and Mojave Deserts, has clarified the complex interrelation between courtship and territoriality that has evolved in this arid-adapted species. Several males may perch on the same creosote bush (*Larrea tridentata*), which is the host plant of their species, while other bushes remain unoccupied (Greenfield and Shelly 1985). Aggregated males mate more often than single males, but not all of them are active in signalling to attract females. Inactive behaviour appears to be a reaction to aggression between males. In spite of this, some of the non-signalling males do achieve reproductive success. Inactive males became active upon moving to a vacant bush, while active males instantly cease their stridulation after moving to a bush containing other signalling males (Shelly and Greenfield 1985). In each reproductive season, the males that mature first tend to settle in the shrubs most heavily occupied during previous years, rather than in those on which their maturation has taken place or in those utilized by their parental generation (Greenfield and Shelly 1987). Females are not evenly distributed among the male mating territories: they appear to select bushes on the basis of the chemical quality of the plant (Shelly et al. 1987). Double phonotaxis has been described in the Texas bush katydid *Scudderia texensis* (Spooner 1964). The male produces two different calls: one of these attracts females from a long distance, the second stimulates females to emit another acoustic signal which, in turn, attracts the male.

Males are vocal and females silent in almost all anurans. The sounds produced by males attract conspecific females during egg-laying. Both sexes of the North American toad *Bufo boreas*, however, are silent (Schmidt 1970) while, among some terrestrial leptodactylids, the females call and attract males to their subterranean nests (Jameson 1981). In desert reptiles, intraspecific communication by sound is common only among the geckos (Bellairs 1959). For example, Namibian barking geckos of the genus *Ptenopus* show pronounced vocal ability (Haacke 1969). The small garrulous gecko *Ptenopus garrulus*, which lives in sandy areas of the Kalahari Desert, emerges from its hole at sunset and produces a characteristic chirping until nightfall (Bellairs 1969). The web-footed *Palmatogecko rangei*, which inhabits loose sand dunes in the Namib Desert, produces a curious and characteristic evening song. Even among desert skinks there are species, such as the Turkestan *Teratoscincus scincus*, which are able to stridulate like grasshoppers (Bellairs 1969). The exact function of the acoustic signals of these reptiles is as yet unknown: they appear most likely to be dependent on territorial habits, but may also be connected with reproductive behaviour (Bellairs 1969).

Birds are the most accomplished singers in the animal kingdom. Their songs are generally associated with courtship and territoriality but, among species living in wide areas with sparse populations, their function is to attract mates (Jameson 1981). In arid environments, the problem as to how male and female birds are able to meet has unfortunately not been extensively investigated. Nevertheless, among various nomad species of the Australian deserts, it has been ascertained that the gathering of conspecific individuals is conditioned by the appearance of abundant food resources (Davies 1984).

Among mammals, attraction between the sexes is often achieved by olfactory or visual signals. Nevertheless, in certain species, such as the mara (or Patagonian cavy) *Dolichotis patagonum*, acoustic interactions between the sexes have been described (Dubost and Genest 1973). Another rodent species, the North American desert grasshopper mouse (*Onychomys leucogaster*), is endowed with a rich vocal repertoire whose function is as yet unclear (Hafner and Hafner 1979). Among African elephants, *Loxodonta africana*, the females in mid-oestrous produce loud, low frequency calls that attract distant males and incite intermale competition (Poole 1989). In the Saharan fennec *Fennecus zerda*, a mating call, which assists in bringing the partners together, is made by both sexes (Gauthier-Pilters 1962).

In other arid-adapted animals, meeting of the sexes depends on visual signals exclusively. For instance, in the Sicilian mutillid *Smicromyrme viduata*, males fly low above the area reconnoitered by the wingless females in search of nests of the sand wasp *Bembex mediterranea* (Sect. 8.5). The

Fig. 53. Mating of the mutillid *Smicromyrme viduata*

male, attracted by the gaudy colour of the females, tries to catch one of them. After a nuptial flight, mating takes place (Fig. 53).

8.3 Courtship and Mating

Courtship is a behavioural system of intraspecific communication, the function of which is the achievement of 'peak productivity' in sexual reproduction, thus maximizing the chance for survival of the species from generation to generation. According to some authors, it may also include the long-distance attraction of the sexes which was discussed in the preceding section.

A fundamental role of courtship is to guarantee the reproductive isolation of the species. It consists of a specific concatenate series of signals and appropriate answers which act in turn as further signals. Only the precise features and temporal sequence of this communicative system result in successful mating. A single difference in the courtship messages of sympatric sibling species is sometimes enough to prevent interspecific hybridization. For example, cadence of sound is the only differing feature in the mating calls of the North American spadefoot toads *Scaphiopus hammondi* and *S. bombifrons* (Blair 1955).

In addition to environmental factors (Sect. 8.1), courtship itself can act as synchronizer for the two sexes. Among insects, such as orthopterans and beetles, pheromones alone may induce reproductive synchrony. In the desert locust *Schistocerca gregaria*, for instance, sexual maturation of both sexes is hastened by the presence of mature males (Norris 1954). The stimulus provided by a courting male appears to be an essential prerequisite to egg-laying by the female of the iguanid lizard *Anolis carolinensis* (Fox 1958). In the Australian budgerigar *Melopsittacus undulatus*, specific precopulatory male vocalization stimulates ovarian development and further song stimulates egg-laying (Brockway 1965).

The determining influence of male sexual pheromones on female ovulation has been demonstrated among different species of rodents. The oestrous cycle of the desert pocket mouse *Perognathus penicillatus* is induced by the presence of the male (Ostwald et al. 1972). In social birds, there is also the possibility of multiple synchronization being induced by the first courting pair (Fraser Darling effect). Mating sometimes occurs after a complex courting procedure in which both partners are engaged. This is the case with the North American grasshopper mouse *Onychomys leucogaster*, whose courtship comprises several phases and can last up to 3 h (Ruffer 1965). Soon after meeting, the male and female circle each other a few times. The female then sniffs the male's genital region. Another bout of circling precedes nose to nose contact, after the two have risen on their hind legs. Courtship may be interrupted and resumed again. After a 'nose kiss', the male sniffs the female's genitalia. He then parades back and forth in front of the female who remains sitting on her haunches. At this point, the male rises on his hind legs and grooms the female's face. Then he alternates between sniffing, grooming, and rubbing his back against her belly. After another prolonged period of courtship, the male succeeds in stimulating the female. Lying on one side, to assume the mating posture, he grooms her neck and other side. A kind of embrace then follows, the two partners grasping each other with their forelegs. This precedes the adoption of a lordosis posture by the female. Finally, the pair rolls over onto their sides and mating takes place (Ruffer 1965).

Another determining role of courtship is that of sexual selection. According to Darwin (1874), this can occur in two different forms: intrasexual selection among courting competitors (usually males) and intersexual selection among mate choosers (usually females). A good example of the involvement of both forms of sexual selection is provided by the reproductive behaviour of the field cricket *Gryllus bimaculatus*, which has been studied in a scrubby Spanish coastal plain (Simmons 1988). Males form calling aggregations. Stridulating males, capable of maintaining individual territories of least 2 m diameter, exploit the best opportunity to attract females, in a similar manner as the natterjack toad *Bufo calamita* (Arak 1983). On the other hand, the call songs of males show individual variations, in both intensity and timing, which are strictly related to body size. Simmons (1988) demonstrated that female choice is strongly dependent on the features of the call song. A marked preference is shown for the sound emissions of the largest males.

Intermale competition does not necessarily take place in the presence of potential mates, nor does it involve actual fighting between competitors. Behavioural patterns, such as long-distance mate attractant signals (Sect. 8.2), territory patrols (Sect. 9.2) and hierarchical social organizations, may be factors of sexual selection (see also Bastock 1967). In addition, active female choice often determines which of a number of courting males is to be successful. For example, females of the Australian frog *Uperoleia laevigata* can recognize and choose, among all calling males, those that are about 70% of their own body weight (Robertson 1990). This kind of choice appears to be highly adaptive, since heavier males hamper oviposition and lighter males do not have the quantity of sperm necessary to fertilize the whole clutch. In other anuran species, such as the South African painted reed frog *Hyperolius marmoratus*, there are strong female preferences at low male densities for the low frequency calls emitted by the largest males (Passmore and Telford 1983). As the density of males increases, however, mating becomes random with respect to call frequency and body size (Dyson 1985). Overlapping calls may effectively reduce a female's ability to discriminate between them (Dyson and Passmore 1988).

In many other non-desert and desert animals, the female chooses from the largest mates. For example, male African elephants, *Loxodonta africana*, continue to grow until late in life (Laws 1966), and the sexual preferences of oestrous females result in their mating with males which are old, vigorous and healthy (Poole 1989). Kin recognition by individual male odour appears to affect mate choice among rodents. For example, oestrous females of the North American white-footed mouse *Peromyscus leucopus* show preference for males at a degree of relatedness intermediate between inbreeding and outbreeding (Keane 1990). It is well known that inbreeding causes an increase in homozygosity, and therefore a genetic depression due either to the unmasking of deleterious recessives or to a reduction in the number of heterotic loci. On the other hand, mating between individuals of low genetic similarity can produce a depression either by a disruption of local adaptation or a breakup of co-adapted gene complexes (Templeton 1986).

The primary role of courtship is to persuade potential partners to mate. The prolongation of courtship, as well as the success of the courting sex (generally the male) depends mainly on the relative states of sexual readiness of the partners when they meet (Ewer 1968b). Male courtship may be very brief, consisting of the pursuit or drawing the female, followed by an attempt to persuade her to adopt the mating posture. In some animals, such as camelids, this objective may be attained by violent means. The male dromedary *Camelus dromedarius* uses his neck to push the female down; the guanaco *Lama guanicoe*, in addition, bites at the female's legs to force her to copulate (Pilters 1956). In the southern African meerkat *Suricata suricatta*, a neck grip may be used to subjugate recalcitrant females (Ewer 1963). Another aggressive feature of the courtship of this viverrid is the nose bite, inflicted in such a manner as to avoid serious hurt to the partner. In another southern African viverrid, the dwarf mongoose *Helogale undulata*, the neck grip does not seem to occur prior to mating, but it is used by males during copulation (Zannier 1965). The neck grip has been described as a preliminary stage of the courtship of male mustelids such as the striped skunk *Mephitis mephitis* (Wight 1931).

In other species, courtship requires more time and prolonged procedures. In the kudu, *Tragelaphus strepsiceros*, courting activity may be divided into three phases (Walther 1964a). Initially, there is a conflict of activity between the partners. Mutual sniffing and male flehmen in response to urination by the female may engender aggression if a male approaches too close or too fast. The male reacts by assuming a threatening pose. The second phase is reached when the female becomes more tolerant of the male's presence and permits him to stand beside her with his head on her hump. The conclusive phase is reached after other skirmishes when the female adopts a mating position. The male then mounts, his neck in contact with the female's back. A similar copulatory position is observed in other mammals, including the African elephant (Buss and Smith 1966). In the Thomson's gazelle *Gazella thomsoni* (Walther 1964b), there is a basic similarity in the successive phases, but the courtship procedure is even more elaborate. Body contact is reduced, and visual signals prevail. Initially, the female flees; the male, following her, performs a characteristic display, alternating head-high and head-horizontal postures. The female urinates, and the male replies with flehmen. Then, beating on the ground with his forefeet, he produces a characteristic 'drum roll'. The male now follows close behind the female and synchronizes his gait with hers. Before copulation, he beats one forefoot against one of the female's legs. This may happen while the pair is walking slowly. During mating, the male holds his head high, limiting contact with the female's body to a minimum. Similar leg beating and mating positions have been described in other Artiodactyla such as the giraffe *Giraffa camelopardalis* (Innis 1958) and the Uganda kob *Adenota kob* (Buechner and Schloeth 1965).

In some invertebrates and vertebrates, courtship is based upon a male's gift of food to the female. The gift sometimes has the function of damping or swaying the female's aggression. In other cases, it is directly connected with creating a suitable nutritional state for mating (see Sect. 8.5). Courtship feeding has been reported mainly among spiders, insects and birds. The female of the Australian bush cricket *Requena verticalis* feeds on the male's elaborate spermatophore. This is very rich in protein and may equal 40% of the body weight of the male (Gwynne 1983). In the laboratory, Gwynne (1984) showed that feeding on the spermatophore enhances the reproductive success of the female by increasing both the number and size of the eggs produced. A spermatophore is a common nuptial food gift among Orthoptera (Sakaluk 1986).

Courtship feeding is frequent among birds (Lack 1940). The gift is usually offered before mating. In the North American deserts, however, the male roadrunner *Geococcyx californianus* catches a mouse or baby rat and shows it as though a promise to a female: however, he gives it to her only after the two have copulated (Calder 1967).

8.4 Pair Stabilization; Sexual Groups

In many animal species, the sexual partners only meet for mating (Fig. 54) which, at most, can last for 1 h (brachygamy, according to Selander 1972). In vertebrates, some crustaceans and insects, on the other hand, there is more or less prolonged pair stabilization (Wickler 1976; Hendrichs 1978). Living together is obligatory when parental care demands the involvement of both sexes. It is suitable adaptation to arid environments where the initial meeting of the sexes is difficult, risky or expensive. It is also necessary for co-operative hunting.

Fig. 54. In many orthopterans, such as the Fuegian grasshopper *Bufonacris bruchii*, maintenance of the pair is limited to mating period

Pair bonding is often associated with monogamous habits and with the necessity for biparental care. An example of strong family cohesiveness is furnished in arid habitats by the desert isopods *Hemilepistus aphganicus* (Schneider 1971, 1975) and *H. reaumuri* (Linsenmayr and Linsenmayr 1971; Linsenmayr 1972, 1979). In the Negev Desert, the female *H. reaumuri* usually digs a shallow burrow before mating takes place (Shachak 1980). After mating, both partners co-operate in extending the burrow and, later, in foraging on behalf of the young. When the young wood lice become nutritionally independent, the parents limit their activity to guarding the burrow against intruders and sealing the entrance with their body. They also contribute in this manner to regulation of the humidity within the burrow. Another case of pair stabilization, which involves both partners in burrow excavation, has been described in the Palaearctic carabid beetle *Scarites buparius*, which lives in sandy coastal areas of Sicily (Alicata et al. 1980). In the evening, soon after mating (Fig. 55), the female selects a suitable site and begins digging, while the male removes the sand from the tunnel at short intervals and guards the burrow entrance. This kind of co-operation, which is unique among carabid beetles, is necessary to achieve, in the space of a single night, the required depth for egg deposition.

Among birds, the formation of lasting pairs is almost general, since 91% of all species are monogamous (Wilson 1975). Nevertheless, among many

Fig. 55. Mating of the Palaearctic carabid beetle *Scarites buparius*

monogamous birds, such as passerines, the pair bond may be limited to a single season because the female is not bound to the male's territory in successive years (Verner and Willson 1969; Dorst 1971). A permanent pair bond has been described in some anatids, cranes, penguins and falcons (Lanyon 1963). In the yellow-eyed penguin *Megadyptes antipodes*, for example, pair fidelity can even last 11 years (Dorst 1971). Lifetime pair bonding has been described in several Australian grass finches (Immelmann 1965; Butterfield 1970; Zann 1977). In the zebra finch *Taenopygia guttata*, the male's courtship song is continued well after the end of the reproductive period and functions to maintain the pair bond (Morris 1954; Immelmann 1969). Miller (1979) demonstrated in the laboratory that the female zebra finch shows clear preferences for her mate's song compared with the song of another male even after being separated from him for a long period. In contrast, the union of sexes persists only during the mating period of birds such as the prairie chicken *Tympanuchus cupido* (Lanyon 1963). In this species, parental care is the exclusive domain of the female.

Pair stabilization may often be based on mutual advantages in hunting. It occurs frequently among Carnivora. For example, the black-backed jackal *Canis mesomelas* lives in permanent pairs which hunt together all year-round (Van der Merwe 1953). Prolonged pair bonding is very advantageous when environmental resources, such as sites for nest building or areas with trophic sources, are scarce or scattered irregularly (Snow 1961; Lack 1968). Kirk's dik-dik, *Madoqua kirki*, is a typical arid-adapted species, which lives in permanent pairs (Hendrichs and Hendrichs 1971; Walther 1990). Each pair occupies and defends its territory for years, or even for life. This dwarf antelope requires selected food, browsing preferentially on buds and the more proteinaceous parts of vegetables. Monogamous habits are present in many other mammal species (Kleiman 1977). For example, monogamy occurs among cricetid rodents such as deer mice (*Peromyscus californicus* and *Peromyscus maniculatus*), beach mice (*Peromyscus polionotus*), cactus mice (*Peromyscus eremicus*), and grasshopper mice (*Onychomys leucogaster* and *Onychomys torridus*) (Howard 1949; Mc Cabe and Blanchard 1950; Blair 1951; Egoscue 1960; Eisenberg 1962, 1963, 1968; Ruffer 1965; Horner and Taylor 1968; Pinter 1970). It is also found among Carnivora, such as coyotes (*Canis latrans*), golden jackals (*Canis aureus*), foxes (*Vulpes macrotis*, *Urocyon cineroargenteus*, *Alopex lagopus*), hunting dogs (*Lycaon pictus*), dwarf mongooses (*Helogale parvula*), suricates (*Suricata suricatta*) (Egoscue 1962; Gauthier-Pilters 1962, 1966; Ewer 1963; Kuhme 1965; Van Lawick-Goodall 1971; Rasa 1973; Ryden 1974; Chesemore 1975; Trapp and Hallberg 1975), and other species.

Many monogamous animals are monomorphic, but there are instances, for example among birds of prey, of mating pairs living together. In the sparrow hawk *Falco sparverius*, the female weighs about one-third more than the male. This size dimorphism appears to be associated with different roles in parental care. The larger sex is deputed to provide protection, warmth and food to the young (Kevles 1986). A similar situation has been described among mammals (Ralls 1976). The female is the larger sex in many species belonging to different animal groups including invertebrates, amphibians, reptiles, birds and mammals. Some animals are monomorphic but not monogamous. A striking example is furnished by spotted hyenas, *Crocuta crocuta*, the young males and females of which seem to be identical in external appearance. Even the primary sexual characters are indistinguishable since the females, endowed with overabundant male hormones, have developed pseudo-penises and scrota. This gives birth to the popular idea that each individual can act as both male and female. This species actually lives in large, female-dominated polygynous groups (Kruuk 1972).

Polygamy is widespread in the animal kingdom. Indeed, according to Wilson (1975), it is the basic nuptial system among animals, especially those living in dry habitats (Crook 1964). In contrast, monogamy is a more recent condition, and a response to exceptional environmental pressure. Polygyny is more common because male reproductive success is often a simple function of the number of females inseminated (Alcock 1989). On the other hand, a female can succeed better in increasing her genetic fitness by entering the rich territory of a polygynic male than as the single partner of a monogamous male in poor territory (Verner 1965; Orians 1969). In the Sonoran Desert iguanid lizard *Urosaurus ornatus*, three or four females may live together on the same mesquite tree. Each male controls two adjacent trees and succeeds in mating with six to eight females. Experimental removal of the females causes abandonment of the territory by the male (M'Closkey et al. 1987). When females do not live in groups, they may be attracted individually by a rich territory and may later mate with the owner. This occurs in the South American vicuna *Vicugna vicugna* (Franklin 1974) and the African impala *Aepyceros melampus* (Jarman 1979).

Harems are frequent among other mammals, especially ungulates such as zebras (Klingel 1965, 1967, 1968), camelids (Koford 1957; Gauthier-Pilters 1959, 1974; Franklin 1973, 1974), pronghorn antelopes (Buechner 1950; Bromley 1969), kobs (Kiley-Worthington 1965; Hanks et al. 1969), wildebeest (Estes 1966, 1969), hartebeest (Gosling 1974), gerenuks (Eisenberg 1966), saiga antelopes (Eisenberg 1966), and musk oxen (Tener 1965). A peculiar kind of harem is that of ostriches (*Struthio camelus*) (Sauer and Sauer 1959; Bertram 1980, 1992). At the onset of the mating season, the male takes possession of a wide territory. Here, he determines the position of the future nest. He then selects a dominant female which will be the favoured partner in the first mating. Later, other females may join the pair at their breeding site. One at a time, the subordinate females succeed in achieving copulation with the resident male. In the Neotropical ratite *Rhea americana*, there is a specific polygamous system (Faust 1960; Bruning 1973, 1974). Up to 15 females may temporarily live together and mate with a single male, thereby forming a provisional harem. The male builds a communal nest, in which each female leaves her eggs to be guarded and hatched by him. After a while, the females look for another suitor. At the end of the reproductive season, each female has had the opportunity to mate with up to seven different males. The rhea or nandu therefore, is both polygynous and polyandrous, and this kind of polyandry is called 'sequential'.

Polyandry is much less common than polygyny. It occurs sporadically in both invertebrates and vertebrates, mainly among birds (Jenni 1974) and often in connection with paternal care (Ridley 1978). In some tinamous, such as the brushland tinamou *Nothoprocta cinerascens* (Lancaster 1964), there is a reproductive system similar to that of the nandu; that is, polygyny combined with sequential polyandry. A particular kind of simultaneous and permanent polyandry is present in the Tasmanian hen *Tribonyx mortieri*, whose sexual groups include one female and two males (Ridpath 1972). The males are brothers and, although one of them dominates the other, executing two-thirds of copulations, both co-operate with the female in incubating the eggs and caring for the nestlings. Wife and territory sharing by two brothers are considered to be a case of kin selection (Maynard Smith and Ridpath 1972).

8.5 Parental Care and Developmental Regulation

Parental post-reproductive activities, devoted to increasing the probability of brood survival (thus guaranteeing a successful conclusion to the reproductive effort), can usually be distinguished as assistance and care. The first term refers to a kind of indirect care which is present in oviparous species in which there is no overlap between succeeding generations. It consists of different behavioural patterns, such as building holes, shelters, nests, cocoons, hoarding trophic resources, or selecting suitable oviposition sites. Some authors do not distinguish between assistance and care (Skutch 1976). On the other hand, among several oviparous animals, initial assistance may be followed by egg care and later also by care of the young.

In all environments, arid habitats included, the primary factor in guaranteeing hatching and developmental success is the nutritional state of the female (Riechert 1979). This is a further explanation of seasonal reproduction (Sect. 8.1), as well as of a female's selection of a territorial male able to defend an area rich in food. It may also explain to some extent the origin of nuptial food gifts (Sect. 8.3). In certain species, there is no overlap between the trophic period and the reproductive season. This occurs among crickets of the genus *Brachytrupes* (Caltabiano et al. 1982a; Costa et al. 1987; Costa and Petralia 1990), whose neo-adults ingest and hoard large quantities of food in their burrows prior to hibernation (Sect. 7.1.1). The following spring, their only activity will be reproduction. The nutritional states of their mates depend, therefore, on the quantity of food ingested some months before.

Selection of optimal sites for egg-laying or nesting is usually determined by the female. It is fundamental to the survival of the eggs and young. The desert locust *Schistocerca gregaria*, whose eggs require moist conditions for development (Hunter-Jones 1964), oviposits only when these conditions are available (Norris 1968). Exposure of the eggs of many desert birds to solar radiation is fatal; so they nest only in shaded sites, in caves or in holes (Russell et al. 1972). There are various methods of egg protection. In addition to microclimatic requirements, anti-predatory and trophic factors must be taken into account.

Oviparous *r* strategists lay large numbers of eggs. Among terrestrial animals, this form of parental investment occurs among both invertebrates and vertebrates. Nevertheless, in many species, the number of eggs laid varies according to biological and environmental parameters. In desert snails, clutch size is highly variable from site to site and, moreover, can be influenced by biotic factors, such as life span, population density and whether the species is oviparous or ovoviviparous (Hunter 1968; Yom-Tov 1971, 1972; Heller 1979). In the North American millipede *Orthoporus ornatus*, the number of eggs per clutch ranges from 100 to 500. It depends on the size of the female (Crawford and Matlack 1979). The clutch of the desert isopod *Hemilepistus reaumuri* may contain 30 to 150 eggs (Shachak 1980). In the South African solifugid *Solpuga caffra* some 200 eggs are produced (Lawrence 1949), while, in the case of the Mexican dune spider *Diguetia imperiosa*, the number of eggs ranges up to 1657 (Bentzien 1973). A striking example of extraordinary fecundity among univoltine arthropods is provided by an Australian moth, *Abantiades magnificus*, whose females lay over 18 000 eggs which have a very high mortality (Common 1970).

Single or clustered eggs may be laid in separated sites. This represents another antipredatory strategy. A well-known example is that of the sand wasp *Ammophila campestris*, which lays each egg in a different nest, dug in the soil. Later, the parent provides a suitable quantity of food for each larva (Baerends 1941). Among sand wasps of the genus *Bembex*, there is a particular behavioural pattern consisting of digging a system of false nests around the true one, close to the systems of other individuals (Fig. 56). The Sicilian dune-inhabiting species *Bembex mediterranea* constructs four false nests which make the location of its larvae by the parasitic mutillid *Smicromyrme viduata* more difficult to determine (Alicata et al. 1974, 1982). Other parental antipredatory strategies range from homocromous colouration of egg shells, as in plovers (*Charadrius* spp.), whose eggs are hard to recognize among the pebbles scattered on sandy beaches to chemical defense, as in tortoises. These sprinkle their stinking urine on the ground to keep predators away from their eggs (Cloudsley-Thompson 1982).

Egg care may occur in numerous ways. Active guarding, often connected with defensive, diversionary or aggressive behaviour against predators, is widespread in the animal kingdom. A striking diversionary strategy

Fig. 56. The Sicilian sand wasp *Bembex mediterranea* resorts to a cunning strategy to protect its larvae. *Above* The individual system of false nests; *below* a set of individual systems

of ground-nesting birds was mentioned in Section 5.2.5. In some arachnids, there is no particular care of the eggs, but the newly hatched young remain on their mother's back until after the first moult. This occurs among scorpions (Cloudsley-Thompson 1951) and in the North American vinegaroon *Mastigoproctus giganteus* (Weygoldt 1972). Egg incubation, although typical of birds, is also present in some lizards, including the iguanids *Crotaphytus collaris* (Clark 1964) and *Sceloporus orcutti* (Mayhew 1963b), and the agamid *Amphibolurus barbatus* (Bustard 1966). Incubation also takes place in pythons (Hutchison et al. 1966; Van Mierop and Barnard 1978; Jameson 1981). Among birds, incubation is generally a maternal activity, but there are cases in which it is carried out by both partners alternately, or only by the male. In ostriches (*Struthio camelus*), the dominant hen incubates during the day, the male during the night. There is, therefore, well-defined co-operation between the two partners; but, determination of the number of eggs to be incubated is the female's prerogative. Depending on the size of the sexual groups of this species (Sauer and Sauer 1959; Sect. 5.4), other females may copulate with the male and lay their eggs close to those laid by the dominant female. However, the latter is able to recognize her own eggs and usually discards most of the others, abandoning them to insolation and predators (Bertram 1979). In Antarctica, the male king penguin *Aptenodytes patagonicus* and the emperor penguin *Aptenodytes forsteri* (Prevost 1961, 1965; Le Maho 1977) incubate the single egg in a fold of skin on their feet for about 50–60 days. Care of the young is carried out by both parents alternately. In the Australian emu *Dromaius novaehollandiae* (Davies 1968) and the South American nandu *Rhea americana* (Bruning 1973; see also Sect. 8.4), egg and chick care is almost exclusively paternal. Paternal care has been described among some primates, such as the wild Japanese macaque, *Macaca fuscata fuscata* (Itani 1959). Vigilance and defensive strategies were mentioned in Section 5.2.7.

Developmental success can also be influenced by parental efforts (e.g. the mother's nutrional state, suitable nest-site selection, and/or efficient parental assistance), as well as by specific morphological, physiological and behavioural brood adaptations. In an unpredictable environment, such as the desert, these factors are particularly important (Noy-Meir 1974; Riechert 1979). First, morphophysiological features allow a wide tolerance of thermal stress. This adaptation is clearly evident among animals, such as crustaceans and amphibians, which develop in temporary rain pools. Unusual temperature tolerance has evolved in the egg of the Pacific tree frog *Hyla regilla* (Schechtman and Olson 1941) and the western spadefoot toad *Scaphiopus hammondi* (Brown 1967). In both species, the range of tolerance increases as development proceeds. The embryos of *Scaphiopus hammondi* can tolerate water temperature of 39 °C in their later stages (Brown 1967; Zweifel 1968), while the dry eggs of the tadpole shrimp *Triops granarius* can tolerate temperatures up to 98 °C for 16 h and may therefore be killed only by boiling (Carlisle 1968)!

The rates of development of the eggs, embryos and young of desert animals are faster than those of animals living in non-arid environments. This has been shown to be the case in several animals, belonging to different groups, such as crustaceans (Rzòrska 1961; Cole 1968), insects (Husain 1937; Hamilton 1950; Hodgkin and Watson 1958; Hunter-Jones 1970; Stinner et al. 1974), amphibians (Brown 1967; Mayhew 1968; Zweifel 1968), reptiles (Orr 1974), birds (Lowe and Hinds 1969; Ohmart 1973; Ponomareva and Grazhdankin 1973) and mammals (Butterworth 1961; Happold 1970; Van der Graaf 1973).

When environmental conditions are too severe, embryonic development may be suspended through quiescence or dormancy. Among invertebrates, suspended development has been described in nematodes (Mankau et al. 1973), mites (Wallace 1970, 1971), crustaceans (Buxton 1923; Rzòrska 1961; Cole 1968) and insects (Shulov and Pener 1963; Jaeger 1965; Wallace 1968; Huque and Juleel 1970).

In the desert roadrunner *Geococcyx californianus* (Ohmart 1973), asynchronous production of eggs may be a behavioural adaptation that regulates clutch size according to environmental conditions. After hatching, the parents supply most of the food to the older young. The smaller nestlings may even be eaten by their parents in times of food shortage. In other species, such as the nandu (Faust 1960), although the eggs are laid over a period of 10 or more days, hatching is almost contemporaneous. This synchronization appears to be induced by sound communication between the embryos (Vince 1969).

In ovoviviparous species, the fertilized eggs are retained within the oviduct until just prior to hatching, but without any trophic relation between the mother and embryos. Among desert animals such bodily protection of the brood is present in lizards, including the California limbless lizard *Anniella pulchra* (Coe and Kunkel 1905; Miller 1944), the North American horned iguanid lizard *Phrynosoma douglassi* (Stone and Rehn 1903; Dammann 1949), and snakes such as the South African puff adder *Bitis arietans* (Fitzsimons 1962; Rose 1962), various species of the genus *Crotalus* (Klauber 1936, 1944, 1956; Fautin 1946; Werler 1951), and some subspecies of the Afro-Asian saw-scaled viper *Echis carinatus* (Stemmler-Gyger 1965; Minton 1966). Ovoviviparity may be considered to be the first step to viviparity, which involves direct trophic contribution by the mother to her embryos. The female of the yucca lizard *Xantusia vigilis* nourishes two embryos through a chorioallantoic placenta (Cowles 1944). A rudimentary placenta has been described in some other reptiles, such as skinks (Robertson et al. 1965; Greer 1977) and North American garter snake *Thamnophis sirtalis* (Neill 1964). Cases of alloparental care will be discussed in the next chapter.

9 Social Behaviour

Social behaviour is widespread in the animal kingdom. It consists of different forms of intraspecific interaction, based upon more or less specialized interindividual co-operation over more or less prolonged periods. This interaction can range from the minimal co-operation of a pair in mating to the more elaborated behavioural patterns of true societies (Tinbergen 1953). Among insects (Wilson 1971, 1975), which include up to 12 000 social species, it is possible to distinguish between 'eusociality', characteristic of the true societies of ants, bees, wasps and termites, and 'presociality'. The latter may include 'subsociality', when parents care for their nymphs or larvae, and 'parasociality', when there are one or two of the following three behavioural patterns: co-operation in parental care, division of reproductive labour, and overlap of the generations that contribute to social activity. The contemporaneous presence of all three patterns corresponds to the eusocial condition. Other terms employed in describing the social organization of insects are: 'community life', when there is a co-operation in nesting but parental care is an individual problem; 'quasisociality', when individuals belonging to the same generation share the same nest, breed and co-operate in brood care; and 'semisociality', when individuals belonging to the same generation co-operate in brood care but play different roles in reproductive labour. (Some individuals are primarily reproductive forms, others mainly workers.) Eusociality has recently been discovered in the Japanese homopteran aphid *Colophina clematis* (Aoki 1977) and even in a mammal, the African naked mole rat *Heterocephalus glaber* (Jarvis 1981).

9.1 Sociality in Desert Environments

In desert environments, sociality has been investigated among a few groups of insects, especially termites and ants. An apparent advantage of social organization to these arid-adapted forms is high microclimatic stability in the communal nest (Crawford 1981). Some examples were mentioned in Section 4.5. In the Negev Desert, survival of the social desert isopod *Hemilepistus reaumuri*, whose family organization includes both parents and up to about

150 juveniles, depends strictly on generalized co-operation in digging an elaborate burrow system in the soil (Shachak 1980). The female constructs an initial burrow which can reach a depth of about 10 cm, while the male alternates his activity between guarding the colony and removing faeces from the burrow. Later, the young combine to extend the family burrow both downward and laterally, constructing several tunnels and chambers at a depth of about 50 cm. To reach such depths, which are indispensable for maintaining suitable moist conditions during periods of drought, it is imperative that all members of the family are involved. The cohesion is guaranteed by family-specific pheromonal badges (Linsenmayr 1972; Shachak et al. 1979).

Another evident advantage conferred by sociality in an unpredictable environment is the possibility for long-term storage of food. Foraging activity is strongly influenced by the abundance and quality of food and/or climatic factors, which vary greatly from season to season and from year to year. This has been demonstrated in various North American seed-harvesting ants (Whitford and Ettershank 1975; Rissing and Wheeler 1976; Whitford 1978b) and termites (Johnson and Whitford 1975; Ueckert et al. 1976). Carnivorous animals profit from communal hunting, by increasing the number of successes and by subduing larger prey than they could kill individually (Curio 1976). Communal hunting and group raiding are typical of ants that form the largest colonies (Wilson 1971). At the same time, co-operative hunting is common among social animals, such as birds of prey (e.g. Brown and Amadon 1968; Berndt 1970; Brauning and Lichtner 1970), wild dogs and jackals (Van Lawick-Goodall and Van Lawick-Goodall 1970), lions and spotted hyenas (Eloff 1964; Kruuk 1972), etc.

Social life is also advantageous in providing protection against predators. In open areas that afford insufficient shelter and vegetative cover, socially organized individuals have increased chances of survival. Alarm signals are widespread among social insects (Wilson 1971). They are nearly always based upon chemical communication but, among non-desert termites, acoustic communication may be involved (Howse 1964). Alarm calls are frequently employed by vertebrates (Sect. 5.2.7). Among territorial animals such as orthopterans, anurans and birds, whose sound production may be almost continuous, the sudden interruption of signalling can function as an alarm signal. For example, some Sonoran Desert katydids are able to intercept the ultrasounds emitted by predatory bats (Spangler 1984). Calling from the tops of shrubs of *Larrea tridentata*, they abruptly stop their song and by their silence inform their conspecifics of the proximity of an enemy. Since it may occur among non-kin individuals, alarm signalling has been interpreted as reciprocal altruism (Trivers 1971). Different cases of alloparental co-operation are also known. For example, unrelated individuals of the dwarf mongoose *Helogale parvula*, which have recently joined established family groups, take active part in social activities, acting as helpers in providing food for the young and in defending the den against predators (Rood 1978). The meaning of this form of altruism is as yet uncertain (Krebs and Davies 1981).

Other social defense strategies, such as the alternation of vigilance activity and carrying out anti-predatory group formations, were mentioned in Section 5.2.7.

Interspecific co-operation has been described in different animal groups. In the Mojave Desert, for instance, up to 200 individuals, belonging to several different families of birds, form large mixed flocks which exploit the available food resources according to a well-coordinated pattern (Cody 1971). In the African plains, impressive mixed herds of impala, wildebeest, hartebeest, gazelles, zebras, giraffes, wart hogs and baboons, move and feed together. Moreover, each species profits by the alarm signals of the others (Washburn and De Vore 1961; Altmann and Altmann 1970; Elder and Elder 1970).

9.2 Territorial Phenomena

Territoriality is, perhaps, the main type of social interaction between animals and their environment. Very few species are truly nomadic, since most animals form more or less close relationships, even if for limited periods, with particular territories having suitable features. An example of very brief territorial behaviour is that of the brindled gnu *Connochaetes taurinus* (Estes 1969). Males of this wildebeest species cease their nomadic migration only for a few hours, or at most a few days, to occupy and defend individual reproductive territories, within which each enacts its courting display.

Territory is a valuable resource, which is advantageous when the benefits it confers overcome the costs of its defense (Brown 1964). Territory must not be so large that the resident spends too much time in its defense, thus profiting little at great cost. Moreover, the energy invested in territorial behaviour involves warning displays and fighting, which may not only cause injury but may make animals more conspicuous to their predators. It can therefore exists only when there is very strong selective pressure. This, for instance, has been shown to be the case among iguanid lizards (Rand 1967). Territoriality is, of course, connected with intraspecific competition through various degrees of agonistic behaviour. In desert environments, where food resources are irregular, competition appears to be less important than elsewhere. Maximum exploitation of the resources (and subsequent maximum brood productivity) may be a more profitable strategy in deserts (Cloudsley-Thompson 1991). In fact, some species, showing territorial habits in non-desert environments, are less sedentary and aggressive in arid areas. This is found, for instance, in the North American brown-shouldered lizard *Uta stansburiana* (Stamps 1977) and in the leopard lizard *Gambelia wislizenii* (Wiens 1976; Cloudsley-Thompson 1991). On the other hand, Tinkle (1967) showed that *Uta stansburiana* may dispense with territorial behaviour under laboratory conditions.

Nevertheless, since territoriality may sometimes confer autoprotective and/or reproductive advantages (Mainardi 1968), various desert animals, belonging to both invertebrate and vertebrate groups, do show territorial habits. For example, burrowing animals such as wood lice (Linsenmayr 1972, 1984; Shachack et al. 1979), arachnids (e.g. Riechert 1978; Cloudsley-Thompson 1991), insects (e.g. Lee and Wood 1971; Wilson 1971; Darlington 1982; Jones and Nutting 1989; Rissing and Pollock 1989), reptiles (e.g. Evans 1961; Carpenter 1967) and mammals (Rood 1970; Nevo 1979) defend their holes against intruders. Several shrub-dwelling animals, such as orthopterans (Alexander 1961; Greenfield and Shelly 1985) and birds, are also territorial. In the North American desert tortoise *Gopherus agassizi* interindividual relations differ according to the seasonal period. Animals spend the winter in dormant groups, but they occupy individual burrows in summer (Woodbury and Hardy 1948). During the reproductive season, males fight until one of the two contenders has been defeated. The loser then emits a particular vocalization as though to ask the winner for help in reassuming a normal position (Patterson 1971). Among birds (Nice 1941; Hinde 1956; Brown 1964; Howard 1964; Fretwell and Lucas 1970) and mammals (Burt 1943; Morris 1965; Mitani and Roadman 1979), territoriality is widespread. The diurnal viverrid meerkat *Suricata suricatta* shows efficient social organization, based upon division of labour among the members of social groups. At the borders of their territory, individuals of this species become extremely pugnacious and severely bite intruders (Ewer 1968b).

Interspecific territorial fighting is well known among desert animals including ants (Wilson 1971), iguanid lizards (Rand 1967), birds (Simmons 1951; Orians and Wilson 1964; Cody 1969; Murray 1971), and rodents (Miller 1964; Ackerman and Weigl 1970; Nevo 1979). According to Wilson (1975), the more similar sympatric species are, the more interspecific competition becomes probable: in fact, the releasers of intraspecific aggressiveness are easily transferred to interspecific relationships.

9.3 Learning and Culture

Since learning appears to be a biological system by which an animal can escape from its stereotype and modify its normal behaviour according to environmental conditions, we expect the desert to be an excellent training place for its inhabitants. Unfortunately, this subject has not received the attention that it merits. The evolution of learning mechanisms in dry habitats is still known only through a series of isolated studies. I shall confine myself to mention those I consider to be the most significant.

Learning phenomena are known in all behavioural spheres, from self-protection to social life. For example, self-protective strategies, such as mimicry and aposematism, are effective only if predators remember their negative experiences. In an analogous manner, hierarchical organization is based upon the learning of different roles by each member of a social group.

Learning of new abilities is sometimes induced by shortage of alimentary resources. In response to such a problem, an animal can learn to exploit food that differs from its usual diet. It may even attempt to leave its habitat. In the desert environment, there is a marked trend toward opportunistic and generalist feeding habits (Chap. 7). Scorpions, locusts, anurans, lizards, birds of prey, hedgehogs, golden moles, rodents, hares, camels, gazelles, coyotes, jackals, bats, and baboons are only a few examples. Moreover, lack of moisture induces several animals to modify their diet in an appropriate manner, or to look for hidden water sources. For example, during the driest periods, desert baboons, such as *Papio hamadryas* and *Papio anubis* (Kummer 1971) and *Papio ursinus* (Brain 1990), gemsbok (Hamilton et al. 1977) and zebras (Louw and Seely 1982), have learned to find water in dry riverbeds.

The progressive shifting of populations from their former habitat may induce important evolutionary changes not only in diet but also in general behaviour. An interesting example is provided by certain desert lizards. Iguanids of the genus *Uta*, living in the northwestern American deserts, can be considered as being among the best adapted of all vertebrates to terrestrial life (Wilson 1975). However, there is one striking exception: in San Pedro Martir, a desert islet located in the Mexican Gulf, the endemic species *U. palmeri* has become partially marine, feeding on intertidal invertebrates during ebb tides (Soulé 1966). Land resources would be insufficient to support a dense population of these large animals. A further step in this evolutionary process has been carried out by the marine iguana *Amblyrhynchus cristatus* of the Galapagos Islands, which dwells on rocks of lava emerging from the sea, and is able to swim underwater, diving to a depth of 11 m to feed upon seaweed (Carpenter 1966). Males are territorial only during the reproductive period, while females usually lay their eggs randomly. On Hood Island, however, where available areas for nesting are scarce, females have acquired territorial habits, competing for nesting sites on the beaches (Eibl-Eibesfeldt 1966).

Use of tools is observed in the insects, birds and mammals (Alcock 1972). Sometimes, it is part of the innate ethogram of species. For example, solitary wasps of the genus *Ammophila*, which show elaborate parental care (Sect. 8.5), occlude the entrances to their nests with little stones, according to a stereotyped pattern of behaviour (Evans and West Eberhard 1970). Larvae of neuropterans (ant lions) and dipterans of the genera *Vermileo* and *Lampromya* repeatedly flick grains of sand towards their prey, causing it to fall to the bottom of their pitfall traps (Topoff 1977; Cloudsley-Thompson 1982). The Egyptian vulture *Neophron percnopterus* selects big stones and hurls them against the eggs of ostriches to break them (Van Lawick-Goodall and Van Lawick-Goodall 1966). A similar behavioural pattern is shown by the Australian black-breasted buzzard kite *Hamirostra melanosternum* on the eggs of emus and other ground-nesting birds (Millikan and Bowman 1967). Some woodpecker finches of the Galapagos Islands are able to extract insects hidden in tree cracks with the help of sticks or cactus spines held in their beaks (Bowman 1961; Eibl-Eibesfeldt and Sielmann 1962). The seven-stone spider *Ariadna* sp., which lives in the gravel plains of the Namib Desert, most probably utilizes the ring of quartz stones, arranged around the entrance of its hole, to detect vibrations produced by the prey (Costa et al. 1993, 1995). In my opinion, even nuptial gifts may be considered as tools.

In certain instances, use of tools is apparently a product of the learning process. This has been shown to be the case mainly among primates and other vertebrates (Van Lawick-Goodall 1964, 1970). Chimpanzees are generally forest-dwellers; therefore, it may appear unjustified to cite them in a book devoted to arid-adapted animals. However, some chimpanzee populations inhabit the savannah areas of equatorial Africa. Moreover, many scientists believe that an important phase of the evolutionary history of this species took place in a semi-desert environment (e.g. Kortland 1962, 1965; Kortland and Kooij 1963; Crook and Garlan 1966; Fox 1967). Man excluded, chimpanzees appear to have the greatest bent of all animals for using tools (Van Lawick-Goodall 1968). The young must often train themselves before accomplishing correctly various aspects of behaviour. For instance, in angling for termites with sticks, adults select sticks of suitable size and, after removing the lateral branches, carry them to the nearest termitarium. They introduce the modified sticks, one at a time, into a termitarium to 'angle' for the termites that cling with their mandibles to foreign objects. The young pick up discarded sticks and, in their play, imitate the adults. After approximately 2 years of such training, they master this food-procuring technique. In other animals, such as the woodpecker finch *Cactospiza pallida*, the use of tools also requires an initial period of apprenticeship (Eibl-Eibesfeldt 1980).

If the term 'culture' means "the totality of learned and socially transmitted behaviours" (Nicholson 1968), we can readily accept that its employment is not limited to Man. Several animal species are clearly endowed with behavioural traditions which are learned by imitation or even by tuition, and are therefore transferred from generation to generation (Mainardi 1980). Many migratory species follow traditional routes; juveniles learn the route by moving in the company of experienced individuals. Food preferences may be transmitted from parents to their young in various ways. Recognition of enemies, as well as hunting techniques, may also be learned by tradition.

Many predators, especially Carnivora and primates, guide the early hunting experiences of their young. Lion cubs perfect their killing technique, assisted by their mothers (Carr 1962). According to Schenkel (1966), cubs are accepted into the pride from the age of 5 months and can share in the communal kill. The cheetah also trains its cubs for the chase, placing at their disposal a living prey animal, such as a Thomson's gazelle fawn (Kruuk and Turner 1967). Among baboons, the mother initially persuades her young to feed on dead prey; then she places unconscious prey at their disposal, so that they can learn the rudiments of hunting. Finally, she goes hunting with the young (Marais 1969). Such animals are therefore endowed with a kind of 'instinct to teach' (Ewer 1969).

Use of tools may also be transmitted by tradition. This is seen in the 'fishing' for termites by chimpanzees. This ability is not, however, shown by all populations. It is absent, for example, in West African chimpanzees (Albrecht and Dunnett 1971). In an analogous way, the technique of cracking nuts with sticks and stones is employed only by populations that live on the Ivory Coast (Struhsaker and Hunkeler 1971). An intriguing question concerns the origin of using sticks as weapons by chimpanzees. According to Kortland (1965), this pattern of behaviour developed in open savannah where the chance of escaping fast predators, such as leopards, is slight. Later, the use of weapons may have disappeared among populations driven by early hominids into the forests.

Abandoning wooded habitats and the colonization of open areas by our remote ancestors were fundamental in human evolution. The erect stance and bipedal gait disengaged the hands from a locomotory function. The use of the hands could therefore be devoted to other activities, such as tool-using and tool-making. Other consequences of bipedalism include new forms of feeding, reproduction and social habits (Lovejoy 1981), and laid the foundations for the extraordinary cultural evolution of Man. Unfortunately, the most recent progress of our species has not always (and still does not) coincide with a rational and far-seeing exploitation of natural resources. A paradoxical effect of the human cultural evolution has been a crisis of identity (Wylie 1971). The presumption of omnipotence has, step by step, transformed the role of Man from that of tool-maker to destroyer of the environment. One of the results has been the increasing desertification or dryland degradation of the Earth (e.g. Cloudsley-Thompson 1978b; Warren 1984; Seely 1991).

A desertified area is not a desert. The latter is a well-balanced ecosystem, whose arid-adapted biocoenosis has survived in a stable equilibrium from time immemorial. I have endeavoured to demonstrate this in reviewing the behavioural adaptations of desert animals. Desertification is, in contrast, a progressive process of the almost irreversible degradation of a productive environment into a barren, lifeless wasteland. It is, perhaps, only a vain hope that, by overcoming the problems created by Mankind, human survival will depend on a new biological and/or cultural evolution within a wholly desertified world!

References

Able KP (1980) Mechanisms of orientation, navigation and homing. In: Gauthreaux SA Jr (ed) Animal migration, orientation and navigation. Academic Press, New York, pp 284–373

Abushama FT (1984) Epigeal insects. In: Cloudsley-Thompson JL (ed) Sahara Desert. Pergamon Press, Oxford, pp 129–144

Abushama FT, Abdel-Nour HO (1973) Damage inflicted on wood by the termite, *Psammotermes hybostoma* Desneux. Z Angew Entomol 73:216–223

Ackerman R, Weigl PD (1970) Dominance relation of red and grey squirrels. Ecology 51:332–334

Adler K, Phillips JB (1985) Orientation in a desert lizard (*Uma notata*): time-compensated compass movement and polarotaxis. J Comp Physiol 156:547–552

Albrecht H, Dunnett SC (1971) Chimpanzees in western Africa. Piper, Munich

Alcock J (1972) The evolution of the use of tools by feeding animals. Evolution 26: 464–473

Alcock J (1989) Animal behavior, 4th edn. Sinauer, Sunderland

Alexander AJ (1958) On the stridulation of scorpions. Behaviour 12:339–352

Alexander AJ (1960) A note on the evolution of stridulation within the family Scorpionidae. Proc Zool Soc Lond 133:391–399

Alexander RD (1961) Aggressiveness, territoriality, and sexual behavior in field crickets (Orthoptera: Gryllidae). Behaviour 17:130–223

Alicata P, Caruso D, Costa G, Motta S (1974) Ricerche eco-etologiche sulla fauna delle dune costiere di Portopalo (Siracusa). I. *Smicromyrme viduata* (Pall.) (Hymenoptera, Mutillidae): ritmi di attività, migrazioni e accoppiamento. Animalia 1:89–108

Alicata P, Caruso D, Costa G, Marcellino I, Motta S, Petralia A (1979) Ricerche eco-etologiche sulla fauna delle dune costiere di Portopalo (Siracusa). II. Comportamento, distribuzione e ritmi di attività di *Pimelia grossa* Fabr (Coleoptera, Tenebrionidae). Animalia 6:33–48

Alicata P, Caruso D, Costa G, Marcellino I, Motta S, Petralia A (1980) Ricerche eco-etologiche sulla fauna delle dune costiere di Portopalo (Siracusa). III. Comportamento e ritmi di attività di *Scarites buparius* Forst. (Coleoptera, Carabidae). Animalia 7: 5–21

Alicata P, Caruso D, Costa G, Marcellino I, Motta S, Petralia A (1982) Studi eco-etologici su Artropodi delle dune costiere di Portopalo (Siracusa, Sicilia). Quaderni sulla struttura delle zoocenosi terrestri. 3. Ambienti mediterranei. I. Le coste sabbiose. CNR AQ/1/178, Roma, pp 159–183

Allen RD (1959) Ameboid movement. In: Brachet J, Mirsky AE (eds) The cell, vol II. Academic Press, New York, pp 135–216

Allen RD (1961) A new theory of ameboid movement and protoplasmic streaming. Exp Cell Res (Suppl) 8:17–31

Allen RD (1962) Amoeboid movement. Sci Am 206:112–122

Allred DM (1965) Note of phalangids at the Nevada test site. Great Basin Nat 25:37–38

Allred DM, Mulaik S (1965) Two isopods at the Nevada test site. Great Basin Nat 25: 43–47

Altmann SA, Altmann J (1970) Baboon ecology: African field research. Univ Chicago Press, Chicago

Anderson J (1898) Zoology of Egypt, vol I. Reptilia and Batrachia. Bernard Quaritch, London

Aoki S (1977) *Colophina clematis* (Homoptera: Pemphigidae), an aphid species with 'soldiers'. Kontyu, Tokyo 45:276–282

Arak A (1983) Sexual selection by male-male competition in natterjack toad choruses. Nature 306:261–262

Aspland KK (1967) Ecology of the lizards in the relictual Cape Flora, Baja California. Am Midl Nat 77:462–475

Attenborough D (1979) Life on Earth, a natural history. David Attenborough Productions, London, and Rizzoli Editore, Milano

Baerends GP (1941) Fortpflanzungsverhalten und Orientierung der Grabwespe *Ammophila campestris*. Tijdschr Entomol 84:68–275

Bailey VE, Sperry CC (1929) Life history and habits of grasshopper mice, genus *Onychomys*. US Dep Agric Tech Bull 145, Washington DC, pp 1–20

Baker RR (1978) The evolutionary ecology of animal migration. Hodder and Stoughton, London

Barnes RD (1968) Invertebrate zoology. WB Saunders, Philadelphia

Barrett C (1928) Lizards in Australian wilds. Bull NY Zool Soc 21:98–111

Bartholomew GA (1956) Temperature regulation in the macropod marsupial, *Setonix brachyurus*. Physiol Zool 29:26–40

Bartholomew GA (1960) The physiology of desert birds. Anat Rec 137:338

Bartholomew GA, Cade TJ (1956) Water consumption of house finches. Condor 58: 406–412

Bartholomew GA, Cade TJ (1957a) The body temperature of the American kestrel, *Falco sparverius*. Wilson Bull 69:149–154

Bartholomew GA, Cade TJ (1957b) Temperature regulation, hibernation, and aestivation in the little pocket mouse, *Perognathus longimembris*. J Mammal 38:60–72

Bartholomew GA, Cade TJ (1963) The water economy of land birds. Auk 80:504–539

Bartholomew GA, Dawson WR (1954) Body temperature and water requirements in the mourning dove, *Zenaidura macroura marginella*. Ecology 35:181–187

Bartholomew GA, Dawson WR (1968) Temperature regulation in desert mammals. In: Brown GW Jr (ed) Desert biology I. Academic Press, New York: pp 395–421

Bartholomew GA, MacMillen RE (1960) The water requirements of mourning doves and their use of sea water and NaCl solutions. Physiol Zool 33:171–178

Bastock M (1967) Courtship: a zoological study. Heinemann, London

Bauer AM, Russell AP (1991) Pedal specializations in dune-dwelling geckos. J Arid Environ 20:43–62

Beck DE, Allred DM (1968) Faunistic inventory-BYU ecological studies at the Nevada test site. Great Basin Nat 28:132–141

Beier M (1962) Ergebnisse der Zoologischen Nubien-Expedition 1962. III. Pseudoscorpionidea. Ann Naturhist Mus Wien 65:297–303

Bellairs A (1959) Reproduction in lizards and snakes. New Biol 30:73–90

Bellairs A (1969) The life of reptiles. Weindenfeld and Nicolson, London

Bennet EL, Bonner J (1953) Isolation of plant growth inhibitors from *Thamnosma montana*. Am J Bot 40:29–33

Bentzien MM (1973) Biology of the spider *Diguetia imperiosa* (Araneida: Diguetidae). Pan Pac Entomol 49:110–123

Benyon P, Rasa OAE (1989) Do dwarf mongooses have a language?: warning vocalizations transmit complex information. S Afr J Sci 85:447–450

Bequaert JC, Miller WB (1973) The mollusks of the arid southwest. Univ Arizona Press, Tucson

Berndt R (1970) Zur Bestandsentwicklung der Greifvögel (Falconiformes) im Dromling. Beitr Vogelkd 16:1–12

Bertram BCR (1979) Ostriches recognise their own eggs and discard others. Nature 279:233–234

Bertram BCR (1980) Vigilance and group size in ostriches. Anim Behav 28:278–286

Bertram BCR (1992) The ostrich communal nesting system. Princeton Univ Press, Princeton

Binda MG, Pilato G (1987) Tardigradi dell'Africa. V. Notizie sui Tardigradi del Nord Africa e descrizione della nuova specie *Macrobiotus diffusus*. Animalia 14:177–191

Bird RD (1930) Biotic communities of the aspen parklands of central Canada. Ecology 11:356–442

Blair WF (1951) Population structure, social behavior and environmental relations in a natural population of the beach mouse (*Peromyscus polionotus leucocephalus*), Contrib Lab Vertebr Biol Univ Mich 48:1–47

Blair WF (1955) Differentiation of mating call in spadefoots, genus *Scaphiopus*. Tex J Sci 7:183–188

Blair WF (1976) Adaptations of anurans to equivalent desert scrub of North and South America. In: Goodall DW (ed) Evolution of desert biota. Univ Texas Press, Austin, pp 197–222

Blanford WT (1888–1891) Fauna of British India. Mammalia. Taylor & Francis, London

Blest AD (1957) The function of eyespot patterns in the Lepidoptera. Behaviour 11: 209–256

Bodenheimer FS (1934) Studies on the ecology of Palestinian Coleoptera. II. Seasonal and diurnal appearance and activity. Bull Soc R Entomol Egypte 18:211–241

Bogert CM, Martin del Campo (1956) The gila monster and its allies. Bull Am Mus Nat Hist 109:1–238

Bonner JT (1980) The evolution of culture in animals. Princeton Univ Press, Princeton

Bornman CH (1972) *Welwitschia mirabilis*: paradoxe du Désert du Namib. Endeavour (Lond) 31:95–99

Bossert WH, Wilson EO (1963) The analysis of olfactory communication among animals. J Theor Biol 5:443–469

Bostic DL (1966) Food and feeding behavior of the teiid lizard, *Cnemidophorus hyperythrus beldingi*. Herpetologica 22:23–31

Bowman RJ (1961) Morphological differentiation and adaptation in the Galapagos finches. Univ Calif Press, Berkeley

Bradshaw G, Howard B (1960) Mammal skulls recovered form owl pellets in Sonora, Mexico. J Mammal 41:282–283

Bragg AN (1946) Aggregation with cannibalism in tadpoles of *Scaphiopus bombifrons* with some general remarks on the probable evolutionary significance of such phenomena. Herpetologica 3:89–96

Bragg AN (1957) Dimorphism and cannibalism in tadpoles of *Scaphiopus bombifrons* (Amphibia, Salientia). Southwest Nat 1:105–108

Bragg AN (1962) Predator-prey relationship in two species of spadefoot tadpoles with notes on some other features of their behavior. Wasmann J Biol 20:81–97

Bragg AN (1964) Further study of predation and cannibalism in spadefoot tadpoles. Herpetologica 20:17–24

Bragg AN (1965) Gnomes of the night. Univ Pennsylvania Press, Philadelphia

Brain C (1988) Water gathering by baboons in the Namib Desert. S Afr J Sci 84:590–591

Brain C (1990) Aspects of drinking by baboons (*Papio ursinus*) in a desert environment. In: Seely MK (ed) Namib ecology. 25 years of Namib research. Transvaal Mus Monogr, Pretoria 7:169–172

Brattstrom BH (1953) Notes on a population of leaf-nosed snakes *Phyllorhynchus decurtatus perkinsi*. Herpetologica 9:57–64

Brauning C, Lichtner J-C (1970) Gemeinsame Jagd zweier Merline. Vogelwelt 91:32
Brian MV (ed) (1978) Production ecology of ants and termites. IBP, vol 13. Cambridge Univ Press, Cambridge
Brinck P (1956) The food factor in animal desert life. In: Wingstrand WG (ed) Bertil Hanstrom: zoological papers in honour of his 65th birthday, November 20th, 1956. Zool Inst, Lund, pp 120–137
Brindle A (1984) The Esphalmeninae (Dermaptera: Pygidicranidae): a group of Andean and southern African earwigs. Entomol 9:281–292
Bristowe WS (1941) The family of spiders. Bernard Quaritch, London
Bristowe WS (1958) The world of spiders. Collins, London
Brockway BF (1965) Stimulation of ovarian development and egg-laying by male vocalization in budgerigars (*Melopsittacus undulatus*). Anim Behav 13:575–578
Bromley PT (1969) Territoriality in pronghorn bucks on the National Bison Range, Moiese, Montana. J Mammal 50:81–89
Bronson FH (1974) Pheromonal influences on reproductive activities in rodents. In: Birch MC (ed) Pheromones. North-Holland, Amsterdam, pp 344–365
Brown GD (1974) The biology of marsupials of the Australian arid zone. Aust Mammal 1:269–288
Brown HA (1967) High temperature tolerance of the eggs of a desert anuran, *Scaphiopus hammondi*. Copeia: 365–370
Brown JL (1964) The evolution of diversity in avian territorial systems. Wilson Bull 76:160–169
Brown L, Amadon D (1968) Eagles, hawks, and falcons of the world, vols 1, 2. Hamlyn House, Feltham
Brownell PH (1977) Compressional and surface wave in desert sand used by scorpions to locate prey. Science 197:479–482
Brownell PH, Farley RD (1979) Prey-localizing behaviour of the nocturnal desert scorpion, *Paruroctonus mesaensis*: orientation to substrate vibrations. Anim Behav 27:185–193
Brun R (1914) Die Raumorientierung der Ameisen und das Orientierungsproblem im allgemeinen. Gustav Fischer, Jena
Bruning DF (1973) The greater rhea chick and egg delivery route. Nat Hist 82:68–75
Bruning DF (1974) Social structure and reproductive behaviour in the greater rhea. Living Birds 13:251–294
Bryant HC (1911) The horned lizards of California and Nevada of the genera *Phrynosoma* and *Anota*. Univ Calif Publ Zool 9:1–84
Bryant HC (1916) Habits and food of the roadrunner in California. Univ Calif Publ Zool 17:21–50
Buechner HK (1950) Life history, ecology, and range use of the pronghorn antelope in Trans-Pecos Texas. Am Midl Nat 43:257–354
Buechner HF, Schloeth R (1965) Ceremonial mating behaviour in Uganda kob (*Adenota kob thomasi* Neuman). Z Tierpsychol 22:209–225
Bullock TH, Barrett R (1968) Radiant heat reception in snakes. Commun Behav Biol(A) 1:19–29
Burrage BR (1973) Comparative ecology and behaviour of *Chamaeleo pumilus pumilus* (Gmelin) and *C. namaquensis* A. Smith (Sauria: Chamaeleontidae). Ann S Afr Mus 61:1–158
Burt WH (1943) Territory and home range concepts as applied to mammals. J Mammal 24:346–352
Buss IO, Smith NS (1966) Observations on reproduction and breeding behavior of the African elephant. J. Wildl Manage 30:375–388
Bustard HR (1964a) Defensive behavior shown by Australian geckos, genus *Diplodactylus*. Herpetologica 20:198–200
Bustard HR (1964b) The systematic status of the Australian geckos *Gehyra variegata* (Dumeril and Bibron, 1836) and *Gehyra australis* Gray, 1845. Herpetologica 21: 294–302

Bustard HR (1966) Notes on the eggs, incubation and young of the bearded dragon, *Amphibolurus barbatus barbatus* (Cuvier). Br J Herpetol 3:252–259
Butler CG (1970) Chemical communication in insects: behavioral and ecological aspects. In: Johnston J Jr, Moulton DG, Turk A (eds) Advances in chemoreception, vol 1. Communication by chemical signals. Appleton-Century-Crofts, New York, pp 35–78
Butterfield PA (1970) The pair bond in the zebra finch. In: Crook JH (ed) Social behavior in birds and mammals: essays on the social ethology of animals and man. Academic Press, London, pp 249–278
Butterworth BB (1961) A comparative study of growth and development of the kangaroo rats, *Dipodomys deserti* Stephens and *Dipodomys merriami* Mearns. Growth 25: 127–139
Buxton PA (1923) Animal life in deserts. Arnold, London, reprinted in 1995
Cade TJ (1965) Relations between raptors and columbiformes at a desert water hole. Wilson Bull 77:340–345
Cade TJ, Dybas JA Jr (1962) Water economy of the budgerygah. Auk 79:345–364
Calaby JH (1960) A note on the food of Australian desert frogs. W Aust Nat 7:79–80
Calder WA (1967) Breeding behavior of the roadrunner *Geococcyx californianus*. Auk 84:597–598
Calder WA (1968) The diurnal activity of the roadrunner, *Geococcyx californianus*. Condor 70:84–85
Calder WA, Schmidt-Nielsen K (1967) Temperature regulation and evaporation in the pigeon and the roadrunner. Am J Physiol 213:883–889
Caltabiano AM, Costa G, Petralia A (1979a) Ricerche sulla locomozione negli Insetti. III. Il nuoto dei Dermatteri. Animalia 6:67–70
Caltabiano AM, Costa G, Petralia A (1979b) Ricerche sulla locomozione negli Insetti. IV. Deambulazione, nuoto e scavo di *Brachytrupes megacephalus* (Lef) (Orthoptera, Gryllidae). Animalia 6:71–79
Caltabiano AM, Costa G, Petralia A (1981) Ricerche sulla locomozione negli Insetti. V. Il nuoto in alcune specie di Coleotteri Carabidi. Animalia 8:105–114
Caltabiano AM, Costa G, Petralia A (1982a) Ricerche eco-etologiche sulla fauna delle dune costiere di Portopalo (Siracusa). IV. Biologia comportamentale di *Brachytrupes megacephalus* (Lef) (Orthoptera, Gryllidae). Animalia 9:269–292
Caltabiano AM, Costa G, Petralia A (1982b) Il nuoto in Ortotteri Ensiferi. Animalia 9:303–312
Caltabiano AM, Costa G, Petralia A (1983) Ricerche sulla locomozione negli Insetti. VII. Il nuoto in alcune specie di Mantodei. Animalia 10:273–282
Carlisle DB (1968) *Triops* (Entomostraca) eggs killed only by boiling. Science 161: 279–280
Carlisle DB, Ghobrial LI (1968) Food and water requirement of Dorcas gazelle in the Sudan. Mammalia 32:570–576
Carpenter CC (1966) The marine iguana of the Galapagos Islands, its behavior and ecology. Proc Calif Acad Sci (4th Ser) 24:329–376
Carpenter CC (1967) Aggression and social structure in iguanid lizards. In: Milstead WW (ed) Lizard ecology: a symposium. Univ Missouri Press. Columbia, pp 87–105
Carpenter GDH (1949) *Pseudacraea eurytus* (L.) (Lepidoptera, Nymphalidae): a study of a polymorphic mimic in various degrees of speciation. Trans R Entomol Soc Lond 100:71–133
Carr N (1962) Return to the wild, 1965 edn. Fontana Books, London
Caruso D, Costa G (1976) L'apparato stridulatore e l'emissione di suoni in *Armadillo officinalis* Dumeril (Crustacea, Isopoda Oniscoidea). Animalia 3:17–27
Casselton PJ (1984) Breeding birds. In: Cloudsley-Thompson JL (ed) Sahara Desert. Pergamon Press, Oxford, pp 229–240
Cates RG (1981) Host plant predictability and the feeding patterns of monophagous, oligophagous, and polyphagous insect herbivores. Oecologia 48:319–326

Causey NB (1975) Desert millipedes (Spirostreptidae, Spirostreptida) of the southwestern United States and adjacent Mexico. Occas Pap Mus Tex Tech Univ 35:1–12
Chamberlin RV (1943) On Mexican centipedes. Bull Univ Utah 33:55
Chapman G (1950) Of the movement of worms. J Exp Biol 27:29–39
Chapman RF (1962) The ecology and distribution of grasshoppers in Ghana. Proc Zool Soc Lond 139:1–66
Cheke RA (1978) Theoretical rates of increase of gregarious and solitarious populations of the desert locust. Oecologia 35:161–171
Chernov Jr, Striganova BR, Ananjeva SI (1977) Soil fauna of the Polar desert at Cape Cheluskin, Taimyr Peninsula, USSR. Oikos 29:175–179
Chesemore DL (1975) Ecology of the Arctic fox (*Alopex lagopus*) in North America: a review. In: Fox MW (ed) The wild canids. Van Nostrand Reinhold, New York, pp 143–163
Chesler J (1938) Observations on the biology of some South African Acrididae (Orthoptera). Trans R Entomol Soc Lond 87:313–351
Chew RM (1961) Water metabolism of desert-inhabiting vertebrates. Biol Rev Cambridge Philos Soc 36:1–31
Chew RM (1965) Water metabolism of mammals. In: Mayer WV, Van Gelder RG (eds) Physiological mammalogy, vol 2. Academic Press, New York, pp 43–178
Chew RM, Chew AE (1970) Energy relationships of the mammals of a desert shrub (*Larrea tridentata*) community. Ecol Monogr 40:1–20
Chippendale G (1962) Botanical examination of kangaroo stomach contents and cattle rumen contents. Aust J Sci 25:21–22
Chopard L (1938) Les orthoptères desertiques de l'Afrique du Nord. Soc Biogeogr (Paris) 6:220–230
Clark H (1946) Incubation and respiration of eggs of *Crotaphytus* c. *collaris* (Say). Herpetologica 3:136–139
Clark RB (1964) Dynamics in Metazoan evolution. The origin of the coelom and segments. Clarendon Press, Oxford
Cloudsley-Thompson JL (1951) Notes on Arachnida. 16. The behaviour of a scorpion. Entomol Mon Mag 86:105
Cloudsley-Thompson JL (1955) On the function of the pectines of scorpions. Annu Mag Nat Hist 8:556–560
Cloudsley-Thompson JL (1956) Studies in diurnal rhythms. VI. Bioclimatic observations in Tunisia and their significance in relation to the physiology of the fauna, especially woodlice, centipedes, scorpions and beetles. Annu Mag Nat Hist 9:305–328
Cloudsley-Thompson JL (1958) Spiders, scorpions, centipedes and mites. The ecology and natural history of woodlice, 'myriapods' and arachnids. Pergamon Press, London
Cloudsley-Thompson JL (1962) Some aspects of the physiology and behaviour of *Dinothrombium* (Acari). Entomol Exp Appl 5:69–73
Cloudsley-Thompson JL (1964) Terrestrial animals in dry heat: arthropods. In: Dill DB (ed) Adaptation to the environment. Handbook of physiology 4. Am Physiol Soc, Washington DC, pp 451–465
Cloudsley-Thompson JL (1966) Climate and fauna in the central and southern Sudan. Sudan Notes Rec 47:127–136
Cloudsley-Thompson JL (1968) The Merkhiyat Jebels: a desert community. In: Brown GW Jr (ed) Desert biology, vol I. Academic Press, New York, pp 1–20
Cloudsley-Thompson JL (1970) On the biology of the desert tortoise *Testudo sulcata* in Sudan. J Zool (Lond) 160:17–33
Cloudsley-Thompson JL (1974) Physiological thermoregulation in the spurred tortoise (*Testudo graeca* L.). J Nat Hist 8:577–587
Cloudsley-Thompson JL (1975) Terrestrial environments. Croom Helm, London
Cloudsley-Thompson JL (1977a) The desert. GP Putnam's Sons, New York

Cloudsley-Thompson JL (1977b) Diurnal rhythms of locomotory activity in isolated desert locust *Schistocerca gregaria* (Forsk.). J Interdiscipl Cycle Res 8:27–36
Cloudsley-Thompson JL (1977c) Adaptational biology of Solifugae (Solpugida). Bull Br Arachnol Soc 4:61–71
Cloudsley-Thompson JL (1978a) Animal migration. Orbis, London
Cloudsley-Thompson JL (1978b) Human activities and desert expansion. Geogr J 144: 416–423
Cloudsley-Thompson JL (1979) Adaptive functions of the colours of desert animals. J Arid Environ 2:95–104
Cloudsley-Thompson JL (1980) Biological clocks. Their functions in nature. Weidenfeld and Nicolson, London
Cloudsley-Thompson JL (1982) La zanna e l'artiglio. Strategie difensive nel mondo animale. (Original title: Tooth and claw. Defensive strategies in the animal world). Boringhieri, Torino
Cloudsley-Thompson JL (1984) Arachnids. In: Cloudsley-Thompson JL (ed) Sahara Desert. Pergamon Press, Oxford, pp 175–204
Cloudsley-Thompson JL (1986) Evolution and adaptation of terrestrial arthropods. Springer, Berlin Heidelberg New York
Cloudsley-Thompson JL (1990) Thermal ecology and behaviour of *Physadesmia globosa* (Coleoptera: Tenebrionidae) in the Namib Desert. J Arid Environ 19: 317–324
Cloudsley-Thompson JL (1991) Ecophysiology of desert arthropods and reptiles. Springer, Berlin Heidelberg New York
Cloudsley-Thompson JL (1993) The diversity of desert life. Scientific Publ, Jodhpur, India
Cloudsley-Thompson JL, Chadwick MJ (1964) Life in deserts. Dufour Editions, Philadelphia
Cloudsley-Thompson JL, Constantinou C (1983) Stridulatory apparatus of Solifugae (Solpugida). J Arid Environ 7:365–369
Cloudsley-Thompson JL, Crawford CS (1970) Water and temperature relations, and diurnal rhythms of scolopendromorph centipedes. Entomol Exp Appl 13:187–193
Cochran DM (1961) Living amphibians of the world. Doubleday, New York
Cody ML (1969) Convergent characteristics in sympatric species: a possible relation to interspecific competition and aggression. Condor 71:223–239
Cody ML (1971) Finch flocks in the Mohave Desert. Theor Popul Biol 2:142–158
Coe WR, Kunkel BW (1905) The female urogenital organs of the limbless lizard *Anniella*. Anat Anz 26:219–222
Coe WR, Kunkel BW (1906) Studies on the California limbless lizard, *Anniella*. Trans Conn Acad Arts Sci 12:1–55
Coineau Y (1981) Une taupe qui 'nage'. Science et Vie, March:75
Cole GA (1968) Desert limnology. In: Brown GW Jr (ed) Desert biology I. Academic Press, New York, pp 423–486
Common IFB (1970) Lepidoptera (moths and butterflies). In: Waterhouse DF (ed) The insects of Australia. Melbourne Univ Press, CSIRO Melbourne, pp 765–886
Constantinou C, Cloudsley-Thompson JL (1983) Stridulatory structures in scorpions of the families Scorpionidae and Diplocentridae. J Arid Environ 7:359–364
Conti E (1994) L'orientamento in assenza di riferimenti visivi in insetti psammofili. Thesis, Catania University, Catania
Conti E, Costa G, Petralia A (1995) The self-burial behaviour in the Namibian acridid *Acrotylus gracilis* La Greca. Madoqua (in press)
Costa G (1970a) Prime osservazioni sull'orientamento di *Sphingonotus candidus personatus* Zanon (Orthoptera, Acrididae). Boll Acc Gioenia Sci Nat 10:257–263
Costa G (1970b) Osservazioni sull'orientamento di Acrididi viventi su litorali sabbiosi. Boll Zool 37:486

Costa G, Carveni C (1975) Galleggiamento e nuoto in Ortotteri Acrididi. Animalia 2:235–239

Costa G, Messina A (1974) Produzione di suoni in Ortotteri Acrididi psammofili. Boll Zool 41:472

Costa G, Petralia A (1979) Ricerche sulla locomozione degli insetti. I. Il nuoto in *Gryllotalpa quindecim* Baccetti e Capra (Insecta, Orthoptera). Animalia 6:5–9

Costa G, Petralia A (1985) Namib and Sicily: ecological and behavioural convergences in sandy ecosystems. Namib Bull 6:9–11

Costa G, Petralia A (1990) Nuove ricerche sulla biologia comportamentale di *Brachytrupes membranaceus* (Drury) vivente nel Namib Desert nell'area di Gobabeb (Namibia). Boll Acc Gioenia Sci Nat, Catania 23:293–316

Costa G, Leonardi M-E, Petralia A (1982) Ricerche sull'orientamento di *Scarites laevigatus F.* (Coleoptera, Carabidae). I. L'orientamento astronomico. Animalia 9:131–151

Costa G, Leonardi M-E, Petralia A (1983a) Ricerche sull'orientamento di *Scarites laevigatus F.* (Coleoptera, Carabidae). V. L'orientamento in notti di luna piena. Animalia 10:343–357

Costa G, Leonardi M-E, Petralia A (1983b) Ricerche eco-etologiche sulla fauna delle dune costiere di Portopalo (Siracusa). V. Ciclo biologico, ritmi di attività e comporta mento di *Timarcha pimelioides* Schaff. (Coleoptera, Chrysomelidae). Animalia 10: 113–134

Costa G, Leonardi M-E, Petralia A (1987) Reproductive behaviour of the giant cricket *Brachytrupes membranaceus* (Drury) in the Namib. Madoqua 15:217–228

Costa G, Petralia A, Conti E, Hänel C (1995) A 'mathematical' spider living in gravel plains of the Namib Desert. J Arid Environ (in press)

Costa G, Petralia A, Conti E, Hänel C, Seely MK (1993) A seven stone spider on the gravel plains of the Namib Desert. 14th Eur Colloq Arachnol. Boll Acc Gioenia Sci Nat, Catania 26:77–83

Cott HB (1940) Adaptive coloration in animals. Methuen, London, reprinted in 1957

Cowles RB (1944) Parturition in the yucca lizard. Copeia 1944:98–100

Crabill RE (1960) A new American genus of cryptopid centipede, with an annotated key to the scolopendromorph genera from America north of Mexico. Proc US Natl Mus 3:1–15

Crawford CS (1972) Water relations in a desert millipede *Orthoporus ornatus* (Girard) (Spirostreptidae). Comp Biochem Physiol 42A:521–535

Crawford CS (1974) The role of *Orthoporus ornatus* millipedes in a desert ecosystem. US/IBP Desert Biome Research Memorandum, RM 74–34, Logan, Utah, pp 77–88

Crawford CS (1976) Feeding-season production in the desert millipede *Orthoporus ornatus* (Girard) (Diplopoda). Oecologia 24:265–276

Crawford CS (1978) Seasonal water balance in *Orthoporus ornatus*, a desert millipede. Ecology 59:996–1004

Crawford CS (1979) Desert detritivores: a review of life history patterns and trophic roles. J Arid Environ 2:31–42

Crawford CS (1981) Biology of desert invertebrates. Springer, Berlin Heidelberg New York

Crawford CS, Cloudsley-Thompson JL (1971) Water relations and desiccation-avoiding behavior in the vinegaroon *Mastigoproctus giganteus* (Arachnida: Uropygi). Entomol Exp Appl 14:99–106

Crawford CS, Matlack MC (1979) Water relations of desert millipede larvae, larva-containing pellets, and surrounding soil. Pedobiologia 19:48–55

Creighton WS, Creighton MP (1960) The habits of *Pheidole militicida* Wheeler (Hymenoptera: Formicidae). Psyche 66:1–12

Crews D, Garstka W (1982) The ecological physiology of the garter snake. Sci Am 247:158–168

Crook JH (1964) The evolution of social organization and visual communication in the weaver birds (Ploceinae). Behaviour (Suppl) 10:1–178

Crook JH, Garlan JS (1966) Evolution of primate societies. Nature 210:1200–1203

Crovetti A (1970) Risultati delle missioni entomologiche dei Proff G Fiori ed E Mellini nel Nord Africa. XXIV. Note eco-etologiche sulla entomofauna primaverile dello 'Uadi Caàm'. Stud Sassar 18:270–381

Cruxent JM (1968) Theses for meditation on the origin and dispersion of man in South America. In: Biomedical challenges presented by the American Indian. Pan American Health organization, Sci Publ No 165, Washington DC, pp 11–16

Curio E (1976) The ethology of predation. Springer, Berlin Heidelberg New York

Curry-Lindahl K (1962) The irruption of the Norway lemming in Sweden during 1960. J Mammal 3:171–184

Dammann J (1949) Birth of eighteen young *Phrynosoma douglassi* Hernandes. Herpetologica 5:144

Darlington JPEC (1982) The Underground passages and storage pits used in foraging by a nest of the termite *Macrotermes michaelseni* in Kajiado, Kenya. J Zool Lond 198:237–247

Darlington PJ (1938) Experiments on mimicry in Cuba, with suggestions for future study. Trans R Entomol Soc Lond 87:681–695

Darwin C (1859) On the origin of species by means of natural selection or the preservation of favoured races in the struggle for life. Murray, London

Darwin C (1874) The descent of man, and sexual selection in relation to sex. Murray, London

Davey HW (1923) The moloch lizard, *Moloch horridus* Gray. Victoria Nat 40:58–60

Davies SJJF (1968) Aspects of a study of emus in semi-arid western Australia. Proc Ecol Soc Aust 3:160–166

Davies SJJF (1984) Nomadism as a response to desert conditions in Australia. J Arid Environ 7:183–195

Davis DD (1936) Courtship and mating behavior in snakes. Fieldiana Zool 20:257–290

Dawkins R (1978) Replicator selection and the extended phenotype. Z Tierpsychol 47: 61–76

Deacon JE, Minckley WL (1974) Desert fishes. In: Brown GW Jr (ed) II. Desert biology. Academic Press, New York, pp 385–488

Deboutteville CD, Rapoport E (1968) Biologie de l'Amérique Australe, vol V. CNRS, Paris

Dekeyser PL, Derivot J (1959) La vie animale au Sahara. Libraire Armand, Cohin, Paris

Delson J, Whitford WG (1973) Adaptations of the tiger salamander, *Ambystoma tigrinum*, to arid habitats. Comp Biochem Physiol 46A:631–638

Délye G (1968) Recherches sur l'écologie, la physiologie et l'éthologie des fourmis du Sahara. Thesis, Univ Aix-Marseille, Marseille

Den Otter CJ (1974) Setiform sensilla and prey detection in the bird spider *Sericopelma rubronitens auserer* (Araneae, Theraphosidae). Neth J Zool 24:219–235

Devetak ND (1985) Detection of substrate vibrations in the ant-lion larva *Myrmeleon formicarius* (Neuroptera: Myrmeleonidae). Biol Vestn 2:11–22

Dingle H (1980) Ecology and evolution of migration. In: Gauthreaux SA Jr (ed) Animal migration, orientation, and navigation. Academic Press, New York, pp 1–101

Dingman RE, Byers L (1974) Interaction between a fossorial rodent (the pocket gopher, *Thomomys bottae*) and a desert plant community. US/IBP Desert Biome Research Memorandum, RM 74–22, Logan, Utah

Dixon JEW, Louw GN (1978) Seasonal effects on nutrition, reproduction and aspects of thermoregulation in the Namaqua sandgrouse (*Pterocles namaqua*). Madoqua 11: 19–29

Dodge NN (1938) Amphibians and reptiles of Grand Canyon National Park. Grand Canyon Nat Hist Assoc, Nat Hist Bull 9:1–55

Dorst J (1971) The life of birds, vol I. Weidenfeld and Nicolson, London
Dorst J (1974) The life of birds. Weidenfeld and Nicolson, London
Doty RL (1972) Odor preferences of female *Peromyscus maniculatus bairdii* for male odors of *P. m. bairdii* and *P. leucopus noveboracensis* as a function of estrous state. J Comp Physiol Psychol 81:191–197
Doumergue F (1900) Essay on the herpetological fauna of Oran with analytical tables and ideas for the determination of all reptiles and batrachians of Morocco, Algeria and Tunisia. Bull Soc Geogr Archeol Oran 20:89–120; 173–220; 233–296; 343–390
Doyen JT, Somerby R (1974) Phenetic similarity and Mullerian mimicry among darkling beetle (Coleoptera: Tenebrionidae). Can Entomol 106:759–772
Dubost G, Genest H (1973) Le comportement social d'une colonie de Maras, *Dolichotis patagonum*, dans le Parc de Branféré. Z Tierpsychol 35:225–302
Dufour L (1862) Anatomie, physiologie et histoire naturelle des Galéodes. Mem Acad Sci Inst France 17:338–446
Dumortier B (1964) Morphology of sound emission apparatus in Arthropoda. In: Busnel R-G (ed) Acoustic behaviour of animals. Elsevier, Amsterdam, pp 277–345
Dyson ML (1985) Aspects of social behaviour and communication in a caged population of painted reed frogs, *Hyperolius marmoratus*. Thesis, Univ Witwatersrand
Dyson ML, Passmore NI (1988) Two-choice phonotaxis in *Hyperolius marmoratus* (Anura:Hyperoliidae): the effect of temporal variation in presented stimuli. Anim Behav 36:648–652
Eddy TA (1961) Foods and feeding patterns of the collared peccary in southern Arizona. J Wildl Manag 25:248–257
Edmunds M (1974) Defence in animals. A survey of anti-predatory devices. Longman, Harlow
Edney EB, McFarlane J (1974) The effect of temperature on transpiration in the desert cockroach, *Arenivaga investigata*, and in *Periplaneta americana*. Physiol Zool 47:1–12
Edney EB, Franco P, Wood R (1978) The responses of *Arenivaga investigata* (Dictyoptera) to gradients of temperature and humidity in sand studied by tagging with technetium 99M. Physiol Zool 51:241–255
Edney EB, Haynes S, Gibo D (1974) Distribution and activity of the desert cockroach *Arenivaga investigata* (Poliphagidae) in relation to micro-climate. Ecology 55: 420–427
Egoscue HJ (1960) Laboratory and field studies of the northern grasshopper mouse. J Mammal 41:99–110
Egoscue HJ (1962) Ecology and life history of the kit fox in Tooele County, Utah. Ecology 43:481–497
Eibl-Eibesfeldt I (1966) Das Verteidigen der Eiablageplätze bei der Hood-Meerechse (*Amblyrhynchus cristatus venustissimus*). Z Tierpsychol 23:627–631
Eibl-Eibesfeldt I (1980) I fondamenti dell'etologia. Il comportamento degli animali e dell'uomo. Adelphi, Milano
Eibl-Eibesfeldt I, Sielmann H (1962) Beobachtungen am Spechtfinken *Cactospiza pallida* (Sclater und Salvin). J Ornithol 103:92–101
Eisenberg JF (1962) Studies on the behavior of *Peromyscus maniculatus gambelii* and *Peromyscus californicus parasiticus*. Behavior 19:177–207
Eisenberg JF (1963) The intraspecific social behavior of some cricetine rodents of the genus *Peromyscus*. Am Midl Natur 69:240–246
Eisenberg JF (1966) The social organization of mammals. Handb Zool 10(7):1–92
Eisenberg JF (1968) Behavior patterns. In: King JA (ed) Biology of *Peromyscus* (Rodentia). Spec Publ 2, Am Soc Mammal, Stillwater, Oklahoma, pp 451–495
Eisner T, Meinwald J (1966) Defensive secretions of arthropods. Science 153:1341–1350
Elder HB (1956) Watering patterns of some desert game animals. J Wildl Manag 20: 368–378

Elder VH, Elder NL (1970) Social grouping and primate associations of the bushbuck (*Tragelaphus scriptus*). Mammalia 34:356–362
El-Kifl AH, Ghabbour SI (1984) Soil fauna. In: Cloudsley-Thompson JL (ed) Sahara Desert. Pergamon Press, Oxford, pp 91–104
Eloff FC (1963) Observations on the migration and habits of the antelopes of the Kalahari Gemsbok Park, Part IV. Koedoe 5:128–136
Eloff FC (1964) On the predatory habits of lions and hyaenas. Koedoe 7:105–112
Endrody-Younga S (1978) Coleoptera. In: Werger MJA (ed) Biogeography and ecology of southern Africa. W Junk, The Hague, pp 797–821
Enright JT (1961) Lunar orientation of *Orchestoidea corniculata* Stout (Amphipoda). Biol Bull 120:148–156
Estes RD (1966) Behaviour and life history of the wildebeest (*Connochates taurinus* Burchell). Nature 212:999–1000
Estes RD (1969) Territorial behaviour of the wildebeest (*Connochaetes taurinus* Burchell, 1823). Z Tierpsychol 26:284–370
Ettershank G (1971) Some aspects of the ecology and microclimatology of the meat ant, *Iridomyrmex purpureus* (Sm). Proc R Soc Victoria 84:137–151
Evans HE, West Eberhard MJ (1970) The wasps. Univ Mich Press, Ann Arbor
Evans LT (1961) Structure as related to behavior in the organization of populations in reptiles. In: Blair WF (ed) Vertebrate speciation. Univ Texas Press, Austin, pp 148–178
Ewer RF (1963) The behaviour of the meerkat, *Suricata suricatta* (Schreber). Z Tierpsychol 20:570–607
Ewer RF (1968a) A preliminary survey of the behaviour in captivity of the dasyurid marsupial, *Sminthopsis crassicaudata* (Gould). Z Tierpsychol 25:319–365
Ewer RF (1968b) Ethology of mammals. Logos Press, London
Ewer RF (1969) The 'instinct to teach'. Nature 22:698
Faust R (1960) Brutbiologie des Nandus (*Rhea americana*) in Glefangenschaft. Verh Dtsch Zool Ges 42:398–401
Fautin RW (1946) Biotic communities of the northern desert shrub biome in western Utah. Ecol Monogr 16:251–310
Fawcett D (1959) Cilia and flagella. In: Brachet J, Mirsky AE (eds) The cell, vol II. Academic Press, New York, pp 217–298
Field CR (1961) Elephant ecology in the Queen Elizabeth National Park, Uganda. East Afr Wildl J 9:99–123
Fielden LJ, Perrin MR, Hickman GC (1990) Feeding ecology and foraging behaviour of the Namib Desert golden mole, *Eremitalpa granti namibensis* (Chrysochloridae). J Zool Lond 220:367–389
Finley RB (1958) The wood rats of Colorado; distribution and ecology. Univ Kansas Publ, Mus Nat Hist 10:213–552
Fiori G (1956) Risultati delle missioni entomologiche dei Proff G Fiori ed E Mellini nel Nord Africa. XI. Ecologia ed etologia della entomofauna dello 'Uadi Caàm'. Boll Ist Entomol Univ Bologna 22:1–44
Fisher HI (1941) Notes on shrews of the genus *Notiosorex*. J Mammal 22:262–269
Fisher J, Hinde RA (1949) The opening of milk bottles by birds. Br Birds 42: 347–357
Fitzsimons VFM (1935) Scientific results of the Vernay-Lang Kalahari Expedition, March to September 1930. Reptilia and Amphibia. Ann Transvaal Mus 16:295–397
Fitzsimons VFM (1938) Transvaal Museum Expedition to south-west Africa and Little Namaqualand, May to August, 1937. Reptiles and Amphibians. Ann Transvaal Mus 19:153–209
Fitzsimons VFM (1962) Snakes of southern Africa. MacDonald, London
Flake LD (1973) Food habits of four species of rodents on a short-grass prairie in Colorado. J Mammal 54:636–647

Flower SS (1933) Notes on the recent reptiles and amphibians of Egypt, with a list of the species recorded from that kingdom. Proc Zool Soc Lond 3:735–851
Fox R (1967) In the beginning: aspects of hominid behavioural evolution. Man 2:415–433
Fox W (1958) Sexual cycle of the male lizard *Anolis carolinensis*. Copeia 1958:22–29
Fraenkel G (1930) Die Orienterung von *Schistocerca gregaria* zur strahlenden Wärme. Z Vergl Physiol 13:300–313
Fraenkel G, Gunn DL (1940) The orientation of animals. Dover, New York
Franklin WL (1973) High, wild world of the vicuna. Nat Geogr 143:76–91
Franklin WL (1974) The social behaviour of the vicuna. In: Giest V, Walther F (eds) The behaviour of ungulates and its relation to management, vol 1. IUCN Publ, Morges, Switzerland, pp 133–143
Freckman DW (1978) Ecology of anhydrobiotic soil nematodes. In: Crowe JH, Clegg JS (eds) Dried biological systems. Academic Press, London, pp 345–357
Freckman DW, Mankau R (1977) Distribution and trophic structure of nematodes in desert soils. Ecol Bull (Stockholm) 25:511–514
French NR, Stoddard DM, Bobek B (1975) Patterns of demography in small mammal populations. In: Golley FB, Petrusewics K, Ryszkowski L (eds) Small mammals, their reproductivity and population dynamics. IBP, vol 5. Cambridge Univ Press, Cambridge, pp 73–102
Fretwell SD, Lucas HL Jr (1970) On territorial behaviour and other factors influencing habitat distribution in birds. I. Theoretical development. Acta Biotheor 19:16–36
Frisch K von (1949) Die Polarisation des Himmelslichts als orientierender Faktor bei den Tänzen der Bienen. Experientia 5:142–148
Frith HJ, Sharman GB (1964) Breeding in wild populations of the red kangaroo *Megaleia rufa*. CSIRO Wildl Res 9:86–114
Fuller WH (1974) Desert soils. In: Brown GW Jr (ed) Desert biology II. Academic Press, London, pp 31–101
Funk RS (1965) Food of *Crotalus cerastes laterorepens* in Yuma County, Arizona. Herpetologica 21:15–17
Gans C (1952) the functional morphology of the egg-eating adaptations in the snake genus *Dasypeltis*. Zoologica 37:209–243
Gans C (1961) Mimicry in procryptically colored snakes of the genus *Dasypeltis*. Evolution 15:72–91
Gauthier-Pilters H (1959) Einige Beobachtungen zum Droh-, Angriffs- und Kampfverhalten des Dromedarhengstes, sowie über Geburt und Verhaltensentwicklung des Jungtiers, in der nordwestlichen Sahara. Z Tierpsychol 16:593–604
Gauthier-Pilters H (1962) Beobachtungen an Feneks (*Fennecus zerda* Zimm.). Z Tierpsychol 19:440–464
Gauthier-Pilters H (1966) Einige Beobachtungen über das Spielverhalten beim Fenek. Z Säugetierk 31:337–350
Gauthier-Pilters H (1974) The behaviour and ecology of camels in the Sahara, with special reference to nomadism and water management. In: Geist V, Walther F (eds) The behaviour of ungulates and its relation to management, vol 2. IUCN Publ, Morges, Switzerland, pp 542–551
Gehlbach FR (1972) Coral snake mimicry reconsidered: the strategy of self-mimicry. Forma Functio 5:311–320
Gehlbach FR (1974) Evolutionary relations of southwestern ringneck snakes (*Diadophis punctatus*). Herpetologica 30:63–72
George U (1969) Über das Tränken der Jungen und andere Lebensäusserungen des Senegal-Flughuhns, *Pterocles senegallus*, in Marokko. J Ornithol (Leipzig) 110: 181–191
George U (1970) Beobachtungen an *Pterocles senegallus* und *Pterocles coronatus* in der Nordwest-Sahara. J Ornithol (Leipzig) 111:175–188
Gertsch WJ (1949) American spiders. Van Nostrand, New York

Ghabbour SI, Mikhail WSA (1977) Variations in chemical composition of *Heterogamia syriaca* Saussure (Poliphagidae, Dictyoptera); a major component of the Mediterranean coastal desert of Egypt. Rev Biol Ecol Mediterr 4:89–104
Ghabbour SI, Mikhail W, Rizk M (1977) Ecology of soil fauna of Mediterranean desert ecosystems in Egypt. I. Summer populations of soil mesofauna associated with major shrubs in the littoral sand dunes. Rev Ecol Biol Sol 14:429–459
Ghilarov MS (1960) Termites of the USSR, their distribution and importance. Proc New Delhi Symp. UNESCO, Paris, pp 131–135
Ghobrial LI, Nour TA (1975) The physiological adaptations of desert rodents. In: Prakash I, Ghosh PK (eds) Rodents in desert environments. Monogr Biol 28. W Junk, The Hague, pp 413–444
Goddard J (1966) Mating and courtship of the black rhinoceros (*Diceros bicornis* L.). East Afr Wildl J 4:69–75
Gorsuch DM (1934) Life history of the Gambel quail in Arizona. Univ Ariz Bull 5:1–89
Ghosh PK (1975) Thermoregulation and water economy in Indian desert rodents. In: Prakash I, Ghosh PK (eds) Rodents in desert environments. W Junk, The Hague, pp 397–412
Gosling LM (1974) The social behaviour of Coke's hartebeest (*Alcelaphus buselaphus cokei*). In: Geist V, Walther F (eds) The behaviour of ungulates and its relation to management, vol 1. IUCN Publication, Morges, Switzerland, pp 488–511
Gray J (1946) The mechanism of locomotion in snakes. J Exp Biol 23:101–120
Gray J (1961) General principles of vertebrate locomotion. Symp Zool Soc Lond 5: 1–11
Gray J (1968) Animal locomotion. Weidenfeld & Nicolson, London
Gray J, Lissmann HW (1938) Studies in animal locomotion. VII. Locomotory reflexes in the earthworm. J Exp Biol 15:506–517
Gray J, Lissmann HW (1964) The locomotion of nematodes. J Exp Biol 41:135–154
Gray R, Bonner J (1948) An inhibitor of plant growth from the leaves of *Encelia farinosa*. Am J Bot 35:52–57
Greenberg B (1943) Social behaviour of the western banded gecko, *Coleonyx variegatus* Baird. Physiol Zool 16:110–122
Greene HW, McDiarmid RW (1981) Coral snake mimicry: does it occur? Science 213: 1207–1212
Greene HW, Pyburn WF (1973) Comments on aposematism and mimicry among coral snakes. Biologist 55:144–148
Greenfield MD, Shelly TE (1985) Alternative mating strategies in a desert grasshopper: evidence of density-dependence. Anim Behav 33:1192–1210
Greenfield MD, Shelly TE (1987) Variation in host-plant quality: implications for territoriality in a desert grasshopper. Ecology 68:828–838
Greer AE (1977) The systematics and evolutionary relationships of the scincid lizard genus *Lygosoma*. J Nat Hist 11:515–540
Griffin DR (1952) Bird navigation. Biol Rev 27:359–400
Grigg GC (1973) Some consequences of the shape and orientation of 'magnetic' termite mounds. Aust J Zool 21:231–237
Grzimek M, Grzimek B (1960) A study of the game of the Serengeti Plains. Z Säugetierk 25:1–61
Guillion GW (1960) The ecology of Gambel's quail in Nevada and the arid southwest. Ecology 41:518–536
Gunn DL (1960) The biological background of locust control. Annu Rev Entomol 5: 279–300
Gupta RK (1986) The Thar Desert. In: Evenari M, Noy-Meir I, Goodall DW (eds) Hot deserts and arid shrublands, B. Elsevier, Amsterdam, pp 55–99
Gwynne DT (1983) Male nutritional investment and the evolution of sexual differences in Tettigoniidae and other Orthoptera. In: Gwynne DT, Morris GK (eds) Orthopteran

mating systems: sexual competition in a diverse group of insects. Westview Press, Boulder, pp 337–366
Gwynne DT (1984) Courtship feeding increases female reproductive success in bush-crickets. Nature 307:361–363
Haacke WD (1969) The call of the barking geckos (Gekkonidae, Reptilia). Sci Pap Namib Desert Res Stn 46:83–94
Hadley NF (1972) Desert species and adaptation. Am Sci 60:338–347
Hadley NF, Williams SC (1968) Surface activities of some North American scorpions in relation to feeding. Ecology 49:726–734
Hafez M (1980) Highlights to the termite problem in Egypt. Sociobiol 5:147–153
Hafner MS, Hafner DJ (1979) Vocalization of grasshopper mice (genus *Onychomys*). J Mammal 60:85–94
Hall ER, Linsdale JM (1929) Notes on the life history of the kangaroo mouse (*Microdipodops*). J Mammal 10:298–305
Hamilton AG (1950) Further studies on the relation of humidity and temperature to the development of two species of African locusts – *Locusta migratoria migratorioides* (R. and F.) and *Schistocerca gregaria* (Forsk.). Trans R Entomol Soc Lond, Ser A 101:1–58
Hamilton WD (1964) The genetical evolution of social behaviour. J Theor Biol 7:1–52
Hamilton WJ III (1973) Life's color code. McGraw-Hill, New York
Hamilton WJ III (1975) Coloration and its thermal consequences for diurnal desert insects. In: Hadley NF (ed) Environmental physiology of desert organisms. Dowden Hutchinson & Ross, Stroudsburg, pp 67–89
Hamilton WJ III (1985) Demographic consequences of a food and water shortage to desert chacma baboons, *Papio ursinus*. Int J Primatol 6:451–462
Hamilton WJ III (1986) Namib Desert baboon (*Papio ursinus*) use of food and water resources during a food shortage. Madoqua 14:397–407
Hamilton WJ III, Coetzee CG (1969) Thermoregulatory behaviour of the vegetarian lizard, *Angolosaurus skoogi*, on the vegetationless northern Namib Desert dunes. Sci Pap Namib Desert Res St 47:95–104
Hamilton WJ III, Seely MK (1976) Fog basking by the Namib Desert beetle, *Onymacris unguicularis*. Nature 262:284–285
Hamilton WJ III, Buskirk RE, Buskirk WH (1976) Defense of space and resources by chacma (*Papio ursinus*) baboon troops in African desert and swamp. Ecology 52:1264–1272
Hamilton WJ III, Buskirk RE, Buskirk WH (1977) Intersexual dominance and differential mortality of gemsbok *Oryx gazella* at Namib Desert waterholes. Madoqua 10:5–19
Hanks J, Price MS, Wrangham RW (1969) Some aspects of the ecology and behaviour of the defassa waterbuck (*Kobus defassa*) in Zambia. Mammalia 33:471–494
Hansen HJ (1893) Organs and characters in different orders of arachnids. Entomol Medd 4:136–251
Hansen RM (1974) Dietary of the chuckwalla, *Sauromalus obesus*, determined by dung analysis. Herpetol 30:120–123
Happold DCD (1970) Reproduction and development of the Sudanese jerboa, *Jaculus jaculus butleri* (Rodentia, Dipodidae). J Zool Lond 162:505–515
Hardy R (1944) Some habits of the banded gecko in southwestern Utah. Utah Acad Sci Proc 21:71–73
Harris WV (1970) Termites of the Palearctic region. In: Krishna K, Weesner FM (eds) Biology of termites II. Academic Press, London, pp 295–313
Hartland-Rowe R (1955) Lunar rhythm in the emergence of an Ephemeropteran. Nature 176:657
Hartland-Rowe R (1958) The biology of a tropical mayfly *Pavilla adusta* Navas (Ephemeroptera, Polymitarcidae) with special reference to the lunar rhythm of emergence. Rev Zool Bot Afr 58:185–202

Hartline PH (1971) Physiological basis for detection of sound and vibration in snakes. J Exp Biol 59:349–371

Haverty MI, Nutting WL (1975) Natural wood preferences of desert termites. Ann Entomol Soc Am 68:533–536

Hawbreaker AC (1947) Food and moisture requirements of the Nelson antelope ground squirrel. J Mammal 28:115–125

Hawke SD, Farley RD (1973) Ecology and behavior of the desert burrowing cockroach, *Arenivaga* sp. (Dictyoptera, Poliphagidae). Oecologia 11:263–279

Heape W (1931) Emigration, migration and nomadism. Cambridge Univ Press, Cambridge

Heim de Balsac H (1936) Biogéographie des mammifères et des oiseaux de l'Afrique de Nord. Bull Biol Fr Belg (Suppl) 21:1–446

Heimlich EM, Heimlich MG (1947) A case of cannibalism in captive *Xantusia vigilis*. Herpetologica 3:149–150

Helfner JR (1953) How to know the grasshoppers, cockroaches and their allies. Wm C Brown, Dubuque, Iowa

Heller J (1979) Distribution, hybridization and variation in the Israeli landsnail *Levantina* (Pulmonata: Helicidae). Zool J Linn Soc 67:115–148

Hendrichs H (1978) Die soziale Organisation von Säugetierpopulationen. Säugetierkundl Mitt 26:81–116

Hendrichs H, Hendrichs U (1971) Dikdik und Elefanten Oekologie und Soziologie zweier afrikanischer Huftiere. Piper R & Co., Munich

Hensley MM (1949) Mammal diet of *Heloderma*. Herpetologica 5:152

Henwood K (1975) A field-tested thermoregulation model for two diurnal Namib Desert tenebrionid beetles. Ecology 56:1329–1342

Hetherington TE (1989) Use of vibratory cues for detection of insect prey by the sandswimming lizard *Scincus scincus*. Anim Behav 37:290–297

Hill WCO, Porter A, Bloom RT, Seago J, Southwick MD (1957) Field and laboratory studies on the naked mole rat, *Heterocephalus glaber*. Proc Zool Soc Lond 128: 455–514

Hinde RA (1956) The biological significance of the territories of birds. Ibis 98:340–369

Hinde RA, Fisher J (1952) Further observations on the opening of milk bottles by birds. Br Birds 44:393–396

Hintz HF, Schryver HF, Halbert M (1973) A note on the comparison of digestion by New World camels, sheep and ponies. Anim Prod 16:303–305

Hodgkin EP, Watson JAL (1958) Breeding of dragon flies in temporary waters. Nature 181:1015–1016

Hoff CC (1959) The ecology and distribution of the pseudoscorpions of north-central New Mexico. Univ New Mexico Press, Albuquerque

Hoffmann G (1983a) The random elements in the systematic search behavior of the desert isopod *Hemilepistus reaumuri*. Behav Ecol Sociobiol 13:81–92

Hoffmann G (1983b) The search behavior of the desert isopod *Hemilepistus reaumuri* as compared with a systematic search. Behav Ecol Sociobiol 13:93–106

Hoffmann RR, Stewart DRM (1972) Grazers or browsers: a classification based on the stomach-structure and feeding habit of East African ruminants. Mammalia 36: 226–240

Hoffmeister DF, Goodpaster WW (1962) Life history of the desert shrew *Notiosorex crawfordi*. southwest Nat 7:236–252

Hofmeyr MD, Louw GN (1987) Thermoregulation, pelage conductance and renal function in the desert-adapted springbok, *Antidorcas marsupialis*. J Arid Environ 13: 137–151

Holdgate MW (1977) Terrestrial ecosystems in the Antarctic. Philos Trans R Soc Lond (B) 279:5–25

Hora SL (1922) Structural modifications in the fish of mountain torrents. Rec Indian Mus 24:31–61

Horch K (1971) An organ for hearing and vibration sense in the ghost crab *Ocypode*. Z Vgl Physiol 73:1–21
Horn HS (1978) Optimal tactics of reproduction and life history. In: Krebs JR, Davies NB (eds) Behavioral ecology, an evolutionary approach. Blackwell, Oxford, pp 411–429
Horner BE, Taylor JM (1968) Growth and reproductive behavior in the southern grasshopper mouse. J Mammal 49:644–660
Hosking WJ (1923) The moloch lizard (*Moloch horridus* Gray). S Aust Nat 4:143–146
Howard E (1964) Territory in bird life. Collins, London
Howard WE (1949) Dispersal, amount of inbreeding, and longevity in a local population of prairie deermice on the George Reserve, southern Michigan. Contrib Lab Vertebr Biol Univ Mich 43:1–50
Howard WE (1960) Innate and environmental dispersal of individual vertebrates. Am Midl Nat 63:152–162
Howell FC (1970) Early man, revised edn. Time-Life Int, Amsterdam
Howse PE (1964) The significance of the sound produced by the termite *Zootermopsis angusticollis* (Hagen). Anim Behav 12:284–300
Hsiao TH, Kirkland RL (1973) Demographic studies of sagebrush insects as functions of various environmental factors. US/IBP Desert Biome Research Memorandum, RM 73–34, Logan, Utah, pp 1–28
Hughes GM (1952) The co-ordination of insect movements. I. The walking movements of insects. J Exp Biol 29:387–402
Hunter WR (1968) Physiological aspects in nonmarine molluscs. In: Drake ET (ed) Evolution and environment. Yale Univ Press, New Haven, pp 83–126
Hunter-Jones P (1964) Egg development in the desert locust (*Schistocerca gregaria* Forsk.) in relation to the availability of water. Proc R Entomol Soc Lond (A) 39: 25–33
Hunter-Jones P (1970) The effect of constant temperature on egg development in the desert locust *Schistocerca gregaria* (Forsk.). Bull Entomol Res 59:707–718
Huque H, Juleel MA (1970) Temperature-induced quiescence in the eggs of the desert locust. J Econ Entomol 63:1398–1400
Husain MA (1937) A summary on investigations on the desert locust, *Schistocerca gregaria*, at Lyallpur during 1934–35. Fourth Int Locust Conf, Cairo, Appendix 31
Hutchison VH, Dowling HG, Vinegar A (1966) Thermoregulation in a brooding female Indian python, *Python molurus bivittatus*. Science 151:694–696
Immelmann K (1963) Drought adaptations in Australian desert birds. Proc 13th Int Ornithol Congr, Ithaca, New York, 1962, pp 649–657
Immelmann K (1965) Australian finches in bush and aviary. Angus and Robertson, Sydney
Immelmann K (1969) Song development in the zebra finch and other estrildid finches. In: Hinde RA (ed) Bird vocalizations. Cambridge Univ Press, London, pp 61–74
Innis AC (1958) The behaviour of the giraffe, *Giraffa camelopardalis*, in the eastern Transvaal. Proc Zool Soc Lond 131:245–278
Iredale T, Troughton EG (1934) A checklist of the mammals recorded from Australia. Mem Aust Mus 6:1–122
Itani J (1959) Paternal care in the wild Japanese monkey, *Macaca fuscata fuscata*. Primates 2:61–93
Jaeckel SGA Jun (1969) Die Molluskem Südamerikas. In: Fittkau EJ, Illies J, Klinge H, Schwabe GH, Sioli H (eds) Biogeography and ecology in South America,II. W Junk, The Hague, pp 794–827
Jaeger EC (1948) Who trims the creosote bush? J Mammal 29:187–188
Jaeger EC (1957) The North American deserts. Stanford Univ Press, Stanford
Jaeger EC (1965) The California Deserts. Stanford Univ Press, Stanford
Jameson DL (1966) Rate of weight loss of three frogs at various temperatures and humidities. Ecology 47:605–613

Jameson EW Jr (1981) Patterns of vertebrate biology. Springer, Berlin Heidelberg New York
Jander R (1975) Ecological aspects of spatial orientation. Annu Rev Ecol Syst 6:171–188
Jarman MV (1979) Impala social behaviour. Territory, hierarchy, mating and use of space. Fortschr Verhalt ensforsch 21:1–92
Jarvis JUM (1981) Eusociality in a mammal: Cooperative breeding in naked mole-rat colonies. Science 212:571–573
Jenni DA (1974) Evolution of polyandry in birds. Am Zool 14:129–144
Johnson CR (1965) The diet of the Pacific fence lizard, *Sceloporus occidentalis occidentalis* (Baird and Girard) from northern California. Herpetologica 21:114–117
Johnson KA, Whitford WG (1975) Foraging ecology and relative importance of subterranean termites in Chihuahuan desert ecosystems. Environ Entomol 4:66–70
Jones R (1973) Emerging picture of Pleistocene Australians. Nature 246:278–281
Jones SC, Nutting WL (1989) Foraging ecology of subterranean termites in the Sonoran Desert. In: Schmidt JO (ed) Special biotic relationships in the arid south west. Univ New Mexico Press, Albuquerque, pp 79–106
Jourdain FCR (1936) The so-called 'injury feigning' in birds. Oologist'Rec 16:25–37; 62–70
Kaestner A (1968) Invertebrate zoology II. (Translated from German by Levi HW and Levi RL). Interscience, New York
Kawai M (1965) Newly acquired precultural behaviour of the natural troop of Japanese monkeys of Koshima islet. Primates 6:1–30
Kawamura S (1954) A new type of action expressed in the feeding behavior of the Japanese monkey in its wild habitat. Organic Evolution 2:10–13
Keane B (1990) The effect of relatedness on reproductive success and mate choice in the white-footed mouse, *Peromyscus leucopus*. Anim Behav 39:264–273
Keast A (1959) Australian birds: their zoogeography and adaptations to an arid continent. In: Keast A, Crocker RL, Christian CS (eds) Biogeography and ecology in Australia. Monogr Biol, vol 8. W Junk, The Hague, pp 89–114
Keast A, Marshall AJ (1954) The influence of drought and rainfall on reproduction in Australian birds. Proc Zool Soc Lond 124:493–499
Kendeigh SC (1961) Animal ecology. Prentice Hall, Englewood Cliffs
Kennedy JS (1939) the behaviour of the desert locust (*Schistocerca gregaria* Forsk.) in an outbreak centre. Trans R Entomol Soc Lond 89:385–542
Kettlewell HBD (1955a) Selection experiments on industrial melanism in the Lepidoptera. Heredity 9:323–342
Kettlewell HBD (1955b) Recognition of appropriate background by the pale and black phases of Lepidoptera. Nature 175:943
Kettlewell HBD (1973) The evolution of melanism. Clarendon Press, Oxford
Kevles B (1986) Females of the species. Sex and survival in the animal kingdom. Harvard Univ Press, Cambridge
Key KHL (1933) Preliminary ecological notes on the Acrididae of the Cape Peninsula. S Afr J Sci 27:406–413
Kheirallah AM (1979) The ecology of the isopod *Periscyphis granai* (Arcangeli) in the western highlands of Saudi Arabia. J Arid Environ 2:51–59
Kiley-Worthington M (1965) the waterbuck (*Kobus defassa*) Ruppell 1835 and *K. ellipsiprimnus* Ogilby 1833) in East Africa: spatial distribution: a study of the sexual behaviour. Mammalia 29:176–204
Klauber LM (1936) *Crotalus mitchelli*, the speckled rattlesnake. Trans San Diego Soc Nat Hist 8:149–184
Klauber LM (1940) The worm snakes of the genus *Leptotyphlops* in the United States and northern Mexico. Trans San Diego Soc Nat Hist 9:87–162
Klauber LM (1941) The long-nosed snakes of the genus *Rhinocheilus*. Trans San Diego Soc Nat Hist 9:289–332

Klauber LM (1944) The sidewinder, *Crotalus cerastes*, with descriptions of new subspecies. Trans San Diego Soc Nat Hist 10:91–126
Klauber LM (1947) Classification and ranges of the gopher snakes of the genus *Pituophis* in the western United States. Bull Zool Soc San Diego 22:7–81
Klauber LM (1951) The shovel-nosed snake, *Chinoactis*, with descriptions of two new subspecies. Trans San Diego Soc Nat Hist 11:141–204
Klauber LM (1956) Rattlesnakes. Univ Calif Press, Berkeley
Kleiman D (1977) Monogamy in animals. Q Rev Biol 52:39–63
Klingel H (1965) Notes on the biology of the plains zebra *Equus quagga boehmi* Matschie. East Afr Wildl J 3:86–88
Klingel H (1967) Soziale Organisation und Verhalten freilebender Steppen-zebras. Z Tierpsychol 24:580–624
Klingel H (1968) Soziale Organisation und Verhaltensweisen vom Hartmann- und Bergzebras (*Equus zebra hartmannae* und *E.z. zebra*). Z Tierpsychol 25:76–88
Knight MH, Skinner JD (1981) Thermoregulatory, reproductive and behavioural adaptations of the big eared desert mouse, *Malacothryx typica*, to its arid environment. J Arid Environ 4:137–145
Knipper H, Kevan DK McE (1954) Über Flügelfarbung und Sicheingraben von *Acrotylus junodi* Schulthess (Orth., Acrididae, Oedipodinae). Veröff Überseemuneum Bremen (A) 2:406–413
Koeppen W, Geiger R (1930) Handbuch der Klimatologie, 5 vols. Borntraeger, Berlin
Koford CB (1957) The vicuna and the puna. Ecol Monogr 27:153–219
Kortland A (1962) Chimpanzees in the wild. Sci Am 206:128–138
Kortland A (1965) How do chimpanzees use weapons when fighting leopards. Am Philos Soc Yearbook 1965, pp 327–332
Kortland A, Kooij M (1963) Protohominid behaviour in primates. Symp Zool Soc Lond 10:61–88
Koyama H, Lewis ER, Leverenz EL, Baird RA (1982) Acute seismic sensitivity in the bullfrog ear. Brain Res 250:168–172
Kraus O (1966) Philogenie, Chorologie und Systematik der Odontopygoideen (Diplopoda, Spirostreptomorpha). Abh Senckenb Naturforsh Ges 512:1–143
Krebs JR, Davies NB (1981) An introduction to behavioural ecology. Blackwell, Oxford
Krishna D, Prakash I (1956) Hedgehogs of the desert of Rajasthan, pt 2. Food and feeding habits. J Bombay Nat Hist Soc 53:362–366
Kruuk H (1972) The spotted hyena: a study of predation and social behavior. Univ Chicago Press, London
Kruuk H, Turner M (1967) Comparative notes on predation by lion, leopard, cheetah and wild dog in the serengeti area, East Africa. Mammalia 31:1–27
Kuhme W (1965) Freilandstudien zur Soziologie des Hyaenenhundes (*Lycaon pictus lupinus* Thomas 1902). Z Tierpsychol 22:495–541
Kuhnelt W (1965) Nahrungsbeziehungen innerhalb der Tierwelt der Namibwüste (Südwestafrika). Sitzungsber Österr Akad Wiss, Nath Naturwiss Kl, Abt 1 174: 185–190
Kuhnelt W (1975) Beiträge zur Kenntniss der Nahrungsketten in der Namibwüste (Südwestafrika). Verh Ges Okol 1975:197–210
Kummer H (1968) Social organization of the Hamadryas baboons: a field study. Univ Chicago Press, Chicago
Kummer H (1971) Primate societies: group techniques of ecological adaptation. Aldine-Atherton, Chicago
Lack D (1940) Courtship feeding in birds. Auk 57:169–178
Lack D (1968) Ecological adaptations for breeding birds. Methuen, London
La Greca M (1947) Su una particolare maniera di deambulazione di un Acridide: *Tropidopola cylindrica* (Marsch.). Boll Zool 14:83–104

Lambert MRK (1984) Amphibians and reptiles. In: Cloudsley-Thompson JL (ed) Sahara Desert. Pergamon Press, Oxford, pp 205–227
Lancaster DA (1964) Biology of the brushland Tinamou, *Nothoprocta cinerascens*. Bull Am Mus Nat Hist 127:269–314
Landsborough Thompson A (1926) Problems of bird-migration. Witherby, London
Lane C, Rothschild M (1965) A case of Mullerian mimicry of sound. Proc R Entomol Soc Lond (A) 40:156–158
Langer P (1984) Anatomical and nutritional adaptations of wild herbivores. In: Gilchrist FM, Mackie RI (eds) Herbivore nutrition in the subtropics and tropics. Science Press (Pty), Craighall (South Africa), pp 185–203
Lanyon WE (1963) Biology of birds. Natural History Press, New York
Larmuth J (1984) Microclimates. In: Cloudsley-Thompson JL (ed) Sahara Desert. Pergamon Press, Oxford, pp 57–66
Lawrence RF (1949) Observations on the habits of a female solifuge, *Solpuga caffra* Pocock. Ann Transvaal Mus 21:197–200
Lawrence RF (1959) The sand-dune fauna of the Namib Desert. S Afr J Sci 55:233–239
Lawrence RF (1966) The Myriapoda of the Kruger National Park. Zool Afr 2:225–262
Lawrence RF (1975) The Chilopoda of south west Africa. Cimbebasia, Ser A 4:35–45
Laws RM (1966) Age criteria for the African elephant, *Loxodonta a. africana*. East Afr Wildl J 4:1–37
Lee KE, Wood TG (1971) Termites and soils. Academic Press, London
Lees DR, Creed ER (1975) Industrial melanism in *Biston betularia*: the role of selective predation. J Anim Ecol 44:67–83
Leistner OA (1967) The plant ecology of the southern Kalahari. Mem Bot Surv S Afr 38:1–172
Le Maho Y (1977) The emperor penguin: a strategy to live and breed in the cold. Am Sci 65:680–693
Leopold AS (1962) The desert. Time, New York
Leouffre A (1953) Phénologie des insects du Sud-Oranais. Israel. Ha-mo'atsah Ha-le' umit Le-mekhar Ule-Fituah. Spec Publ 2:325–331
Lewis ER, Narins PM (1985) Do frogs communicate with seismic signals? Science 227:187–189
Lidicker WZ Jr (1975) The role of dispersal in the demography of small mammals. In: Golley FB, Petrusewicz K, Ryszkonwski L (eds) Small mammals: their productivity and population dynamics. Cambridge Univ Press, London, pp 103–128
Ligon JD (1968) The biology of the elf owl, *Micrathene whitneyi*. Misc Publ Mus Zool Univ Mich 136:1–70
Linsenmayr KE (1972) Die Bedeutung familienspezifischer 'Abzeichen' für den Familienzusammenhalt bei der sozialen Wüstenassel *Hemilepistus reaumuri* Audouin u. Savigny (Crustacea, Isopoda, Oniscoidea). Z Tierpsychol 31:131–162
Linsenmayr KE (1979) Untersuchungen zur Soziobiologie der Wüstenassel *Hemilepistus reaumuri* und verwandter Isopodenarten (Isopoda, Oniscoidea): Paarbindung und Evolution der Isogamie. Verh Dtsch Zool Ges 46:60–72
Linsenmayr KE (1984) Comparative studies on the social behaviour of the desert isopod *Hemilepistus reaumuri* and of a *Porcellio* species. Symp Zool Soc Lond 53: 423–453
Linsenmayr KE, Linsenmayr C (1971) Paarbildung und Paarzusammenhalt bei der monogamen Wüstenassel *Hemilepistus reaumuri* (Crustacea, Isopoda, Oniscoidea). Z Tierpsychol 29:134–155
Lissmann HW (1945a) The mechanism of locomotion in gastropod molluscs. I. Kinematics. J Exp Biol 21:58–69
Lissmann HW (1945b) The mechanism of locomotion in gastropod molluscs. II. Kinetics. J Exp Biol 22:37–50

Little EL, Keller JG (1937) Amphibians and reptiles of the Jornada Experimental Range, New Mexico. Copeia 1937:216–222

Loomis RF (1966) Descriptions and records of Mexican Diplopoda. Ann Entomol Soc Am 59:11–27

Louw GN (1972) The role of advective fog in the water economy of certain Namib Desert animals. Symp Zool Soc Lond 31:297–314

Louw GN (1993) Physiological animal ecology. Longman, Harlow

Louw GN, Hamilton WJ III (1972) Physiological and behavioural ecology of the ultrapsammophilous Namib Desert tenebrionid, *Lepidochora argentogrisea*. Madoqua 1:87–95

Louw GN, Holm E (1972) Physiological, morphological and behavioural adaptations in the ultrapsammophilous Namib Desert lizard, *Aporosaura anchietae*. Madoqua 1:67–85

Louw GN, Seely MK (1982) Ecology of desert organisms. Longman, London

Louw GN, Belonje PC, Coetzee HJ (1969) Renal function, respiration, heart rate and thermoregulation in the ostrich (*Struthio camelus*). Sci Pap Namib Desert Res St 42:43–54

Lovejoy CO (1981) The origin of man. Science 211:341–350

Lowe CH (1955) Gambel quail and water supply on Tiburon Island, Sonora, Mexico. Condor 57:244

Lowe CH, Hinds DS (1969) Thermoregulation in desert populations of roadrunners and doves. In: Hoff CC, Riedesel ML (eds) Physiological systems in semiarid environments. Univ New Mexico Press, Albuquerque

Lull RS (1940) Organic evolution. Macmillan, New York

MacArthur RH (1972) Geographical ecology. Harper and Row, New York

MacArthur RH, Wilson EO (1967) The theory of island biogeography. Princeton Univ Press, Princeton

MacMillen RE (1962) The minimum water requirements of mourning doves. Condor 64:165–166

MacMillen RE (1964) Population ecology, water relations, and social behaviour of a southern California semidesert rodent fauna. Univ Calif Publ Zool 71:1–66

MacMillen RE (1965) Aestivation in the cactus mouse, *Peromyscus eremicus*. Comp Biochem Physiol 16:227–248

MacMillen RE, Lee AK (1967) Australian desert mice: Independence of exogenous water. Science 158:383–385

MacMillen RE, Trost CH (1966) Water economy and salt balance in white-winged and Inca doves. Auk 83:441–456

Mahendra BC (1936) Contributions to the bionomics, anatomy, reproduction and development of the Indian house-gecko, *Hemidactylus flaviridis* Ruppel, Part I. Proc Indian Acad Sci 4:250–281

Main AR, Littlejohn MJ, Lee AK (1959) Ecology of Australian frogs. In: Keast A, Crocker RC, Christian CS (eds) Biogeography and ecology in Australia. Monogr Biol, vol 8. W Junk, The Hague, pp 396–411

Main BY (1956) Taxonomy and biology of the genus *Isometroides keyserling* (Scorpionida). Aust J Zool 4:158–164

Main BY (1957) Biology of aganippine trapdoor spiders (Mygalomorphae: Ctenizidae). Aust J Zool 5:402–473

Mainardi D (1968) La scelta sessuale nell'evoluzione della specie. Boringhieri, Torino

Mainardi D (1980) Tradition and the social transmission of behavior in animals. In: Barlow GW, Silverberg J (eds) Sociobiology: beyond nature/nurture. AAAS Symp 35, pp 227–255

Mankau R, Sher SA, Freckman DW (1973) Biology of nematodes in desert ecosystems. US/IBP Desert Biome Research Memorandum, RM 73–7, Logan, Utah, pp 1–22

Manton SM (1952a) The evolution of arthropodan locomotory mechanics. Part 2. General introduction to the locomotory mechanics of Arthropoda. J Linn Soc Zool 42:93–117

Manton SM (1952b) The evolution of arthropodan locomotory mechanics. Part 3. The locomotion of the Chilopoda and Pauropoda. J Linn Soc Zool 42:118–166
Manton SM (1953) Locomotory habits and the evolution of the larger arthropodan groups. Symp Soc Exp Biol 7:339–376
Manton SM (1965) The evolution of arthropodan locomotory mechanics. Part 8. Functional requirements and body design in Chilopoda, together with a comparative account of their skeleto-muscular systems and an appendix on a comparison between burrowing forces of annelids and chilopods and its bearing upon the evolution of the arthropodan haemocoel. J Linn Soc Zool 46:251–483
Manton SM (1968) Terrestrial Arthropoda (II). In: Gray J (ed). Animal locomotion. Weidenfeld and Nicolson, London, pp 333–376
Marais E (1969) The soul of the ape. Anthony Blond, London
Mares MA, Rosenzweig ML (1978) Granivory in North and South American deserts: rodents, birds and ants. Ecology 59:235–241
Marinari A, Vinciguerra MT, Vovlas N, Zullini A (1981) Nematodi delle dune costiere d'Italia. Quaderni sulla struttura delle zoocenosi terrestri. 3. Ambienti mediterranei. I. Le coste sabbiose. CNR, AQ/1/173, 1982, Rome, pp 27–50
Marsh AC, Louw GN, Berry HH (1978) Aspects of renal physiology, nutrition and thermoregulation in the ground squirrel *Xerus inauris*. Madoqua 11:129–135
Marsh B (1982) An ecological study of *Welwitschia mirabilis* and its satellite fauna. Namib Bull 4:3–4
Martens R (1960) The world of amphibians and reptiles. Harrap & co., London
Martin AC, Zim HS, Nelson AL (1951) American wildlife and plants. Dover, New York
Matthews EG (1976) Insect ecology. Univ Queensland Press, St Lucia
Mayhew WW (1963a) Biology of the granite spiny lizard, *Sceloporus orcutti*. Am Midl Nat 69:310–327
Mayhew WW (1963b) Reproduction in the granite spiny lizard *Sceloporus orcutti*. Copeia 1963:144–152
Mayhew WW (1965) Adaptations of the amphibian, *Scaphiopus couchi*, to desert conditions. Am Midl Natur 74:95–109
Mayhew WW (1968) Biology of desert amphibians and reptiles. In: Brown GW Jr (ed) Desert biology I. Academic Press, New York, pp 195–356
Maynard Smith J, Ridpath MG (1972) Wife sharing in the Tasmanian native hen *Tribonyx mortierii*: a case of kin selection? Am Nat 106:447–552
McCabe TT, Blanchard BD (1950) Three species of *Peromyscus*. Rood Assoc, Santa Barbara
McCallan E (1956) Observations on self-burial in *Acrotylus hirtus* Dirsh (Orthoptera, Acrididae). Entomol Mon Mag 92:116–117
McCallan E (1964) Ecology of sand dunes with special reference to the insect communities. In: Davis DHS (ed) Ecological studies in southern Africa. W Junk, The Hague, pp 174–185
McCleary JA (1968) the biology of desert plants. In: GW Brown Jr (ed) Desert biology I. Academic Press, New York, pp 141–194
McGinnies WG (ed) (1968) Deserts of the world. An appraisal of research into their physical and biological environment. Univ Arizona Press, Tucson
McGinnis SM, Voigt WG (1971) Thermoregulation in the desert tortoise *Gopherus agassizii*. Comp Biochem Physiol 40A:119–126
M'Closkey RT, Baia KA, Russell RW (1987) Defense of mates: a territory departure rule for male tree lizards following sex-ratio manipulation. Oecologia 73:28–31
McMichael DF, Iredale T (1959) The land and freshwater Mollusca of Australia. In: Keast A, Crocker RL, Christian CS (eds) Biogeography and ecology of Australia. W Junk, The Hague, pp 224–245
Medvedev GS (1965) Adaptations of leg structures in desert darkling beetles. Entomol Rev 44:473–485

Meester J (1964) Revision of the Chrysochloridae.I. The desert golden mole *Eremitalpa* Roberts. Sci Pap Namib Desert Res Stn 26:1–18

Meigs P (1953) World distribution of arid and semi-arid homoclimates. In: Review of research on arid zone hydrology. UNESCO, Paris 1:203–210

Mellini E (1976a) Risultati delle missioni entomologiche dei Proff G Fiori ed E Mellini nel Nord Africa. XXXII. Attività della entomofauna nelle oasi di Mizda e di el-Ghéria esc-Scerghia in primavera avanzata. Boll Ist Entomol Univ Bologna 33:55–114

Mellini E (1976b) Risultati delle missioni entomologiche dei Proff G Fiori ed E Mellini nel Nord Africa. XXXIII. Etologia degli insetti dello Uadi Sofeggìn ed altri uidiàn della Ghìbla nel mese di maggio. Boll Ist Entomol Univ Bologna 33:115–214

Mendelssohn H, Golani I, Marder U (1971) Agricultural development and the distribution of venomous snakes and snake bite in Israel. In: De Vries A, Kochva E (eds) Toxins of animals and plant origin. Gordon and Breach, London

Mertens R (1960) the world of amphibians and reptiles. McGraw-Hill, New York

Michel R (1969) Etude experimentale des variations de la tendance au vol au cours du vieillissement chez le criquet pélerin *Schistocerca gregaria* (Forsk.). Rev Comp Anim 3:46–65

Millar RP (1972) Reproduction in the rock hyrax (*Procavia capensis*) with special preferences to the seasonal sexual activity in the male. Thesis, Univ Liverpool

Miller AH (1951) An analysis of the distribution of the birds of California. Univ Calif Publ Zool 50:531–644

Miller AH, Stebbins RC (1964) The lives of desert animals in Joshua Tree National Monument. Univ Calif Press, Berkeley

Miller CM (1944) Ecologic relations and adaptations of the limbless lizards of the genus *Anniella*. Ecol Monogr 14:271–289

Miller DB (1979) The acoustic basis of mate recognition by female zebra finches (*Taenopygia guttata*). Anim Behav 27:376–380

Miller L (1932) Notes on the desert tortoise (*Testudo agassizii*). Trans San Diego Soc Nat Hist 7:187–208

Miller L (1955) Notes on the desert tortoise, *Gopherus agassizi*, of California. Copeia 1955:113–118

Miller PL (1972) Swimming in mantids. J Entomol 46:91–97

Miller RS (1964) Ecology and distribution of pocket gophers (Geomyidae) in Colorado. Ecology 45:256–272

Millikan GC, Bowman RI (1967) Observations on Galapagos tool-using finches in captivity. Living Birds 6:23–41

Millot J (1943) Les araignées mangeuses de vertebrés. Bull Soc Zool Fr 68:10–16

Minnich J, Shoemaker VH (1970) Diet, behavior and water turnover in the desert iguana, *Dipsosaurus dorsalis*. Am Midl Nat 84:496–509

Minton SA (1966) A contribution to the herpetology of West Pakistan. Bull Am Mus Nat Hist 134:27–184

Minton SA (1968) Venoms of desert animals. In: Brown GW Jr (ed) Desert biology I. Academic Press, New York, pp 487–516

Mispagel ME (1978) The ecology and bioenergetics of the acridid grasshopper, *Bootettix punctatus*, on creosote bush, *Larrea tridentata*, in the northern Mojave Desert. Ecology 59:779–788

Mitani JC, Rodman PS (1979) Territoriality: the relation of ranging pattern and home range size to defendability, with an analysis of territoriality among primate species. Behav Ecol Sociobiol 5:241–251

Mitchell MJ (1977) Life history strategies of oribatid mites. In: Dindal DL (ed) Biology of oribatid mites. State Univ NY Coll Environ Sci For, Syracuse, pp 65–69

Montanucci RR (1965) Observations on the San Joaquin leopard lizard, *Crotaphytus wislizenii silus* Stejneger. Herpetologica 21:270–283

Montanucci RR (1967) Further studies on leopard lizards, *Crotaphytus wislizenii*. Herpetologica 23:119–126
Moore BP (1974) Pheromones in the termite societies. In: Birch MC (ed) Pheromones. North-Holland, Amsterdam, pp 250–266
Moreau RE (1950) The breeding seasons of African birds. I. Land birds. Ibis 92:223–267
Morris D (1954) The reproductive behaviour of the zebra finch (*Poephila guttata*), with special reference to pseudofemale behaviour and displacement activities. Behaviour 6:271–322
Morris D (1965) The mammals. A guide to the living species. Hodder and Stoughton, London
Mosauer W (1932a) On the locomotion of snakes. Science 76:583–585
Mosauer W (1932b) Adaptive convergence in the sand reptiles of the Sahara and of California. A study in structure and behavior. Copeia 1932:72–78
Mosauer W (1933) Locomotion and diurnal range of *Sonora occipitalis*, *Crotalus cerastes*, and *Crotalus atrox* as seen from their tracks. Copeia: 14–16
Moseley HN (1877) Notes on the structures of several forms of land planarians, with a description of two genera and several new species, and a list of all species at present known. Q J Microsc Sci 17:273–292
Mottram JC (1915) Some observations on pattern-blending with reference to obliterative shading and concealment of outline. Proc Zool Soc Lond 1915:679–692
Muma MH (1966) Feeding behavior of North American Solpugida (Arachnida). Fla Entomol 49:199–216
Murie A (1951) Coyote food habits on a southwestern cattle range. J Mammal 32:291–295
Murray BG (1971) The ecological consequences of interspecific terrirorial behavior in birds. Ecology 52:414–423
Muth A (1977) Thermoregulatory postures and orientation to the sun: a mechanistic evaluation for the zebra-tailed lizard *Callisaurus draconoides*. Copeia 1977:710–720
Nagy B (1959) Das Sicheingraben von *Acrotylus longipes* und *A. insubricus*. Acta Zool Acad Sci Hung 5:369–388
Nagy KA (1973) Behavior, diet and reproduction in a desert lizard, *Sauromalus obesus*. Copeia 1973:93–102
Neill WT (1964) Viviparity in snakes: some ecological and zoogeographical considerations. Am Nat 98:35–55
Nevo E (1979) Adaptive convergence and divergence of subterranean mammals. Annu Rev Ecol Syst 10:269–308
Newlands G (1978) Arachnida (except Acari). In: Werger MJA (ed) Biogeography and ecology of southern Africa II. W Junk, The Hague, pp 685–702
Newsome AE (1965a) The abundance of red kangaroos, *Megaleia rufa* (Desmarest) in central Australia. Aust J Zool 13:269–287
Newsome AE (1965b) The distribution of red kangaroos, *Megaleia rufa* (Desmarest) about sources of persistent food and water in central Australia. Aust J Zool 13:289–299
Newsome AE (1966) The influence of food on breeding in the red kangaroo in central Australia. CSIRO Wildl Res 11:187–196
Newsome AE, Corbett LK (1975) Outbreaks of rodents in semi-arid and arid Australia: causes, preventions, and evolutionary considerations. In: Prakash I, Ghosh PK (eds) Rodents in desert environments. Monogr Biol 28. W Junk, The Hague, pp 117–153
Nice MM (1941) The role of territory in bird life. Am Midl Nat 26:441–487
Nichols UG (1953) Habits of the desert tortoise, *Gopherus agassizii*. Herpetologica 9: 65–69
Nicholson CK (1968) Anthropology and education. Merril, Columbus
Norris KS (1949) Observations on the habits of the horned lizard *Phrynosoma m'calli*. Copeia 1949:176–180

Norris KS (1953) The ecology of the desert iguana *Dipsosaurus dorsalis*. Ecology 34: 265–287
Norris KS (1958) The evolution and systematics of the iguanid genus *Uma* and its relation to the evolution of other North American desert reptiles. Bull Am Mus Nat Hist 114:253–326
Norris MJ (1954) Sexual maturation in the desert locust, *Schistocerca gregaria* (Forsk.), with special reference to the effects of grouping. Anti-Locust Bull 18:1–44
Norris MJ (1968) Laboratory experiments on oviposition responses of the desert locust *Schistocerca gregaria* (Forsk.). Anti-Locust Bull 43:1–47
Norton BE, Smith LB (1975) Response to insect herbivory. US/IBP Desert Biome Res Mem 75–15:6
Nour HO Abd El (1980) The natural durability of building wood and the use of wood preservatives in Sudan. Sociobiol 5:175–182
Noy-Meir I (1973) Desert ecosystems: environment and producers. Annu Rev Ecol Syst 4:25–51
Noy-Meir I (1974) Desert ecosystems: higher trophic levels. Annu Rev Ecol Syst 5: 195–214
Nutting WL, Haverty MI (1976) Seasonal production of alates by five species of termites in an Arizona desert grassland. Sociobiology 2:145–153
Nutting WL, Haverty MI, LaFage MI (1974) Colony characteristics of termites as related to population density and habitat. US/IBP Desert Biome Res Mem, RM 74–33, Logan, Utah
Nutting WL, Haverty MI, LaFage JP (1975) Demography of termite colonies as related to various environmental factors: population dynamics and role in the detritus cycle. US/IBP Desert Biome Res Mem 75–31, Logan, Utah, pp 1–26
Odum EP (1971) Fundamentals of ecology. Saunders, Philadelphia
Odum EP, De la Cruz AA (1963) Detritus as a major component of ecosystems. AIBS Bull (now BioScience) 13:39–40
Ohmart RD (1969) Dual breeding ranges in Cassin sparrow (*Aimophila cassini*). In: Hoff CC, Riedesel ML (eds) Physiological systems in semiarid environments. Univ New Mexico Press, Albuquerque
Ohmart RD (1973) Observations on the breeding adaptations of the road-runner. Condor 75:140–149
Orians GH (1969) On the evolution of mating systems in birds and mammals. Am Nat 103:589–603
Orians GH, Solbrig OT (1977) Degree of convergence of ecosystem characteristics. In: Orians GH, Solbrig OT (eds) Convergent evolution in warm deserts. US/IBP Synthesis Ser 1. Dowden Hutchinson & Ross, Stroudsburg, pp 226–255
Orians GH, Willson MF (1964) Interspecific territories of birds. Ecology 45:736–745
Orians GH, Cates RG, Mares MA, Moldenke A, Neff J, Rhoades DF, Rosenzweig ML, Simpson BB, Schultz JC, Tomoff CS (1977) Resource utilization systems. In: Orians GH, Solbrig OT (eds) Convergent evolution in warm deserts. Dowden Hutchinson & Ross, Stroudsburg, pp 164–224
Orr RT (1970) Animals in migration. Collier MacMillan, London
Orr Y (1974) Adaptation of the small-spotted lizard, *Eremias guttulata*. Isr J Med Sci 10:285–286
Orton G (1954) Dimorphism in larval mouth parts in spadefoots of the *Scaphiopus hammondi* group. Copeia 1954:97–100
Ostwald R, Wilken K, Simons J, Highstone H, Cimino S, Shimondle S (1972) Influences of photoperiod and partial contact on estrus in the desert pocket mouse, *Perognathus penicillatus*. Biol Reprod 7:1–8
Otte D (1970) A comparative study of communicative behaviour in grasshoppers. Misc Publ Mus Zool Univ Mich 141:1–168

Otte D, Joern A (1975) Insect territoriality and its evolution: population studies of desert grasshoppers on creosote bushes. J Anim Ecol 44:29–54

Papi F (1960) Orientation by night: the moon. Cold Spring Harbor Symp Quant Biol 25:475–480

Papi F (1990) Homing phenomena: mechanisms and classifications. Ethol Ecol Evol 2: 3–10

Papi F (1992) General aspects. In: Papi F (ed) Animal homing. Animal Behaviour Series. Chapman & Hall, London, pp 1–18

Papi F, Pardi L (1954) La luna come fattore di orientamento degli animali. Boll Ist Mus Zool Univ Torino 4:1–4

Papi F, Pardi L (1959) Nuovi reperti sull'orientamento lunare di *Talitrus saltator* Montagu (Crustacea, Amphipoda). Z Vgl Physiol 41:583–596

Papi F, Pardi L (1963) On the lunar orientation of sandhopper (Amphipoda, Talitridae). Biol Bull 124:97–105

Papi F, Tongiorgi P (1963) Innate and learned components in the astronomical orientation of wolf spiders. Ergeb Biol 26:259–280

Pardi L (1954) Über die Orientierung von *Tylos latreilli* (Auch. und Sav.) (Isopoda terrestria). Z Tierpsychol 11:175–181

Pardi L (1960) Innate components in the solar orientation of littoral amphipods. Cold Spring Harbor Symp Quant Biol 25:395–401

Pardi L, Papi F (1952) Die Sonne als Kompass bei *Talitrus saltator* Montagu (Amphipoda, Talitridae). Naturwissenschuften 39:262–263

Pardi L, Ercolini A, Ferrara F, Scapini F, Ugolini A (1984) Orientamento zonale solare e magnetico in Crostacei Anfipodi litorali di regioni equatoriali. Atti Acc Naz Lincei Rc (Cl Sci fis mat nat) 76:312–320

Parris R, Child G (1973) The importance of pans to wildlife in the Kalahari and the effect of human settlement on these areas. J S Afr Wildl Manage Assoc 3:1–8

Passmore NI, Telford SR (1983) Random mating by size and age of males in the painted reed frog, *Hyperolius marmoratus*. S Afr J Sci 79:353–355

Patterson RG (1971) Vocalization in the desert tortoise, *Gopherus agassizi*. Thesis, Calif State Univ, Fullerton

Pearson OP (1954) Habits of the lizard *Liolaemus multiformis multiformis* at high altitudes in Southern Peru. Copeia 1954:111–116

Penrith M-L (1975) The species of *Onymacris* Allard (Coleoptera: Tenebrionidae). Cimbebasia (Ser A) 4:47–97

Petter F (1961) Répartition géographique et écologique des rongeurs désertiques du Sahara occidental à l'Iran oriental. Mammalia 25:1–219

Pfluger HJ, Burrows M (1978) Locusts use the same basic motor pattern in swimming and kicking. J Exp Biol 75:81–93

Pianka ER (1970) On r- and K-selection. Am Nat 104:592–597

Pianka ER (1971) Comparative ecology of two lizards. Copeia 1971:129–138

Pianka ER (1982) Observations on the ecology of *Varanus* in the Great Victorian Desert. West Aust Nat 15:37–44

Pianka ER (1985) Some intercontinental comparisons of desert lizards. Natl Geogr Res 1:490–504

Pianka ER (1986) Ecology and natural history of desert lizards. Analysis of the ecological niche and community structure. Princeton Univ Press, Princeton

Pianka ER, Parker WS (1975) Ecology of horned lizard: a review with special reference to *Phrynosoma platyrhinos*. Copeia 1975:141–162

Pianka ER, Pianka HD (1970) The ecology of *Moloch horridus* (Lacertilia: Agamidae) in western Australia. Copeia: 90–103

Pilato G (1982) Descrizione di *Hexapodibius bindae* n.sp. e discussione sulla famiglia Calohypsibiidae (Eutardigrada). Animalia 9:213–226

Pilato G, Catanzaro R, Binda MG (1991) Remarks on some tardigrades of the African fauna with the description of three new species of *Macrobiotus* Schultze 1834. Trop Zool 4:167–178

Pilters H (1956) Das Verhalten der Tylopoden. Handb Zool, Berlin, 8(10), No 27:1–24

Pinter AJ (1970) Reproduction and growth for two species of grasshopper mice (*Onychomys*) in the laboratory. J Mammal 51:236–243

Pocock RI (1941) Fauna of British India. Mammalia. Taylor & Francis, London

Polis GA (1979) Prey and feeding phenology of the desert sand scorpion *Paruroctonus mesaensis* (Scorpionidae: Vaejovidae). J Zool Lond 188:333–346

Polis GA (1980) The effect of cannibalism on the demography and activity of a natural population of desert scorpions. Behav Ecol Sociobiol 7:25–35

Polis GA, Farley RD (1979a) Behavior and ecology of mating in the cannibalistic scorpion, *Paruroctonus mesaensis* Stahnke (Scorpionida: Vaejovidae). J Arachnol 7:33–46

Polis GA, Farley RD (1979b) Characteristics and environmental determinants of natality, growth and maturity in a natural population of the desert scorpion, *Paruroctonus mesaensis* (Scorpionida: Vaejovidae). J Zool Lond 187:517–542

Polis GA, Farley RD (1980) Population biology of a desert scorpion: survivorship, microhabitat and the evolution of life history strategy. Ecology 61:620–629

Polis GA, Sissom WD, McCormick SJ (1981) Predators of scorpions: field data and a review. J Arid Environ 4:309–326

Polunin O (1972) The concise flowers of Europe. Oxford Univ Press, London

Pond AW (1962) The desert world. Nelson, New York

Ponomareva TS, Grazhdankin AV (1973) Adaptatsiya ptentsov pustynnykh ptits k teplovomu vozdeistviyu sredy. Zool Zh 52:1528–1536

Poole JH (1989) Mate guarding, reproductive success and female choice in African elephants. Anim Behav 37:842–849

Pope CH (1960) The reptile world. Knopf, New York

Popov AV, Shuvalov VF (1977) Phonotactic behavior of crickets. J Comp Physiol 119:111–126

Popov GB, Wood TG, Haggis MJ (1984) Insect pests of the Sahara. In: Cloudsley-Thompson JL (ed) Sahara Desert. Pergamon Press, Oxford, pp 145–174

Porter WP, Mitchell JW, Beckman WA, DeWitt CB (1973) Behavioural implications of mechanistic ecology: thermal and behavioural modeling of desert ectotherms and their microenvironment. Oecologia 13:1–54

Pough FH (1988) Mimicry and related phenomena. In: Gans C, Huey RB (eds) Biology of the Reptilia 16 (Ecology B). Defense and life history. Alan R Liss, New York, pp 153–234

Pradhan S (1957) The ecology of arid zone insects excluding locusts and grasshoppers. In: Arid zone research, vol VIII. Hum Anim Ecol, UNESCO, Paris, pp 199–240

Prakash I (1956) Studies on the ecology of the desert hedgehogs. Proc Rajasthan Acad Sci 6:30–42

Prakash I (1959a) Foods of certain insectivores and rodents in captivity. Univ Rajasthan Studies (B) 6a:1–18

Prakash I (1959b) Foods of Indian desert mammals. J Biol Sci 21:100–109

Prakash I (1959c) Food of the Indian false vampire. J Mammal 40:545–547

Prakash I (1962) Ecology of gerbils of the Rajasthan Desert India. Mammalia 26:311–331

Prakash I (1969) Eco-toxicology and control of Indian desert gerbille, *Meriones hurrianae* Jerdon. V. Food preference in the field during monsoon. J Bombay Nat Hist Soc 65:581–589

Prakash I (1971) Breeding season and litter size of Indian desert rodents. Z Angew Zool 58:441–454

Prakash I, Kametkar LR (1969) Body weight, sex and age factors in populations of the northern palm squirrel, *Funambulus pennanti* Wroughton. J Bombay Nat Hist Soc 66:99–115

Prakash I, Jain AP, Purohit KG (1971) A note the breeding and post-natal development of the Indian gerbil *Tatera indica indica* in Rajasthan Desert. Säugetierkd Mitt 19: 375–380

Prevost J (1961) Ecologie du Manchot empereur *Aptenodytes forsteri*. Gray, Hermann, Paris

Prevost J (1965) Ecologie des Manchots antarctiques. In:Van Mieghem J, Van Oye P (eds) Biogeography and ecology in Antarctica. Monogr Biol, vol 15. W Junk, The Hague, pp 551–648

Pugach S, Crawford CS (1978) Seasonal changes in hemolymph amino acids, proteins, and inorganic ions of a desert millipede *Orthoporus ornatus* (Girard) (Diplopoda: Spirostreptidae). Can J Zool 56:1460–1465

Rainey RC (1976) Flight behaviour and features of the atmospheric environment. In: Rainey RC (ed) Insect flight. Symp R Entomol Soc Lond 7:75–112

Rainey RC, Waloff Z, Burnett GF (1957) The behaviour of the red locust (*Nomadacris septemfasciata* Serville) in relation to the topography, meteorology, and vegetation of the Rukwa Rift Valley, Tanganyika. Anti-locust Bull 26:1–86

Ralls K (1976) Mammals in which females are larger than males. Q Rev Biol 51:245–276

Rand AS (1967) Adaptive significance of territoriality in iguanid lizards. In: Milstead WW (ed) Lizard ecology: a symposium. Univ Missouri Press, Columbia, pp 106–115

Rao R (1924) A note on cannibalism in a gecko. J Bombay Nat Hist Soc 30:228

Rasa OAE (1973) Intra-familiar sexual repression in the dwarf mongoose, *Helogale parvula*. Naturwissenschaften 6:303–304

Rasa OAE (1985) Coordinated vigilance in dwarf mongoose family group: the 'watchman's song' hypothesis and the cost of guarding. Ethology 71:340–344

Rasa OAE (1989) The costs and effectiveness of vigilance behaviour in the dwarf mongoose: implications for fitness and optimal group size. Ethol Ecol Evol 1: 265–282

Reeve WL (1952) Taxonomy and distribution of the horned lizards genus *Phrynosoma*. Univ Kans Sci Bull 34:817–960

Reichman OJ (1975) The relation of desert rodent diets to available resources. J Mammal 56:731–751

Reichman OJ (1978) Ecological aspects of the diets of Sonoran Desert rodents. Museum of Northern Arizona Research Paper Series 20, Mus North Ariz Res Cent, Flagstaff Annu Rep

Reichman OJ, Van der Graaff K (1975) Association between ingestion of green vegetation and desert rodents reproduction. J Mammal 56:503–506

Reichman OJ, Prakash I, Roig V (1979) Food selection and consumption. In: Goodall DW, Perry RA (eds) Arid-land ecosystems, vol 1. Cambridge Univ Press, Cambridge, pp 681–716

Rhoades DF (1977) The antiherbivore chemistry of *Larrea*. In: Mabry TJ, Hunziker JH, DiFeo DR Jr (eds) Creosote bush. Biology and chemistry of *Larrea* in new world deserts. US/IBP Synthesis Ser 6. Dowden Hutchinson & Ross, Stroudsburg, pp 135–175

Riazance J, Whitford WG (1974) Studies of wood borers, girdlers and seed predators of mesquite. US/IBP Desert Biome Research Memorandum RM 74–30, Logan, Utah

Ricklefs RE (1974) Energetics of reproduction in birds. In: Paynter RA (ed) Avian energetics. Nuttal Ornithol Club No 15, Cambridge, MA, pp 152–292

Riddle WA, Crawford CS, Zeitone AM (1976) Patterns of hemolymph osmoregulation in three desert arthropods. J Comp Physiol 112:295–305

Ridley M (1978) Paternal care. Anim Behav 26:904–932

Ridpath MG (1972) The Tasmanian native hen, *Tribonyx mortierii*. CSIRO Wildl Res 17:53–90

Riechert SE (1974) The pattern of local web distribution in a desert spider: mechanisms and seasonal variation. J Anim Ecol 43:733–745

Riechert SE (1978) Energy-based territoriality in populations of the desert spider *Agelenopsis aperta* (Gertsch). Symp Zool Soc Lond 42:211–222

Riechert SE (1979) Development and reproduction in desert animals. In: Goodall DW, Perry RA (eds) Arid-land ecosystems, vol 1. Cambridge Univ Press, Cambridge, pp 797–822

Riechert SE, Tracy CR (1975) Thermal balance and prey availability: bases for a model relating web-site characteristics to spider reproductive success. Ecology 56:265–284

Riedesel ML, Cloudsley-Thompson JA, Cloudsley-Thompson JL (1971) Evaporative thermoregulation in turtles. Physiol Zool 44:28–32

Riek EF, Michener CD, Brown WLJ, Taylor RW (1970) Hymenoptera (wasps, bees, ants). In: Waterhouse DF (ed) The insects of Australia. Melbourne Univ Press, CSIRO Melbourne, pp 867–959

Riley CV (1878) Quoted in: Williams CB (1958) Insect migration. Collins, London

Rissing SW, Pollock GB (1989) Behavioral ecology and community organization of desert seed-harvester ants. J Arid Environ 17:167–173

Rissing SW, Wheeler J (1976) Foraging responses of *Veromessor pergandei* to changes in seed production. Pan pac Entomol 52:63–72

Robertson JGM (1990) Female choice increases fertilization success in the Australian frog, *Uperoleia laevigata*. Anim Behav 39:639–645

Robertson LAD, Chapman BM, Chapman RF (1965) Notes on the biology of the lizards *Agama cyanogaster* and *Mabuya striata striata* collected in the Rutkwa valley, southwest Tanganyika. Proc Zool Soc Lond 145:305–320

Robinson MD (1980) (unpubl results, University Simon Bolivar, Caracas, Venezuela) Quoted in: Louw GN, Seely MK (1982) Ecology of desert organisms. Longman, London

Robinson MD, Hughes DA (1978) Observations on the natural history of Peringuey's adder, *Bitis peringueyi* (Boulenger) (Reptilia: Viperidae). Ann Transvaal Mus 31: 190–196

Roffey J (1972) Migration and dispersal in the Australian plague locust. Abstr, 14th Int Cong Entomol (Canberra):155

Rollo CD, Wellington WG (1981) Environmental orientation by terrestrial mollusca with particular reference to homing behaviour. Can J Zool 59:225–239

Rood JP (1970) Ecology and social behavior of the desert cavy (*Microcavia australis*). Am Midl Nat 83:415–454

Rood JP (1978) Dwarf mongoose helpers at the den. Z Tierpsychol 48:277–287

Rose W (1962) The reptiles and amphibians of southern Africa. Maskew Miller, Cape Town

Ross A (1961) Notes on food habits of bats. J Mammal 42:66–71

Roth LE (1958) Ciliary coordination in Protozoa. Exp Cell Res (Suppl) 5:573–585

Roth LM, Willis ER (1961) The biotic associations of cockroaches. Smithson Misc Collect 141:470

Rothschild M (1961) Defensive odours and Mullerian mimicry among insects. Trans R Entomol Soc Lond 113:101–121

Ruelle JE (1978) Isoptera. In: Werger MJA (ed) Biogeography and ecology of southern Africa. W Junk, The Hague, pp 747–762

Ruffer DG (1965) Sexual behaviour of the northern grasshopper mouse (*Onychomys leucogaster*). Anim Behav 13:447–452

Russell SM, Smith EL, Gould P, Austin G (1972) Studies on Sonoran birds. US/IBP Desert Biome Research Memorandum, RM 72–31, Logan, Utah, pp 1–6

Ruthven AG (1907) A collection of reptiles and amphibians from southern New Mexico and Arizona. Bull Am Mus Nat Hist 23:483–604

Ryan RM (1968) Mammals of Deep Canyon, Colorado Desert, California. The Desert Museum, Palm Springs

Ryden H (1974) The 'lone' coyote likes family life. Natl Geogr Mag 146:279–294

Rzòska J (1961) Observations on tropical rainpools and general remarks on temporary waters. Hydrobiologia 17:265–286

Rzòska J (1984) Temporary and other waters. In: Cloudsley-Thompson JL (ed) Sahara Desert. Pergamon Press, Oxford, pp 105–114

Sadlier RMFS (1965) Reproduction in two species of kangaroo *Macropus robustus* and *Megaleia rufa* in the arid Pilbara region of western Australia. Proc Zool Soc Lond 145:239–261

Sakaluk SK (1986) Is courtship feeding by male insects parental investment? Ethology 73:161–166

Santos PF, Whitford WG (1979) (pers. comm. in, Wallwork JA 1982)

Santos PF, DePree E, Whitford WG (1978) Spatial distribution of litter and microarthropods in a Chihuahuan desert ecosystem. J Arid Environ 1:41–48

Sargent TD (1966) Background selection of geometrid and noctuid moths. Science 154:1674–1675

Sauer EGF, Sauer EM (1959) Polygamie beim Südafrikanischen Strauss. Zool Beitr 10:266–285

Saunders DS (1976) Insect clocks. Pergamon Press, Oxford

Schaefer GW (1976) Radar observations of insect flight. In: Rainey RC (ed) Insect flight. Symp R Entomol Soc Lond 7:157–197

Schaller GB (1972) The Serengeti lion: a study of predator-prey relations. Univ Chicago Press, Chicago

Schechtman AM, Olson JB (1941) Unusual temperature tolerance of an amphibian egg (*Hyla regilla*). Ecology 22:409–410

Schenkel R (1966) Play, exploration and territoriality in the wild lion. Symp Zool Soc Lond 18:11–22

Schenkel R, Schenkel-Hulliger L (1969) Ecology and behavior of the black rhinoceros, *Diceros bicornis*. Paul Parey, Hamburg

Schmidt KP, Inger RF (1957) Living reptiles of the world. Doubleday, New York

Schmidt RS (1970) Auditory receptors of two mating call-less anurans. Copeia 1970: 143–147

Schmidt-Nielsen K (1958) The resourcefulness of nature in physiological adaptation to the environment. Physiologist 1:4–20

Schmidt-Nielsen K (1964) Desert animals. Oxford Univ Press, Oxford

Schmidt-Nielsen K, Haines HB (1964) Water balance in a carnivorous desert rodent: the grasshopper mouse. Physiol Zool 37:259–265

Schmidt-Nielsen K, Newsome AE (1962) Water balance in the mulgara (*Dasycercus cristicauda*), a carnivorous desert marsupial. Aust J Biol Sci 15:683–689

Schmidt-Nielsen K, Dawson TJ, Hammel HT, Hinds D, Jackson DC (1965) The jack rabbit. A study in its desert survival. Hvalradets Skrifter Norske Videnskaps. Akad Oslo 48:125–142

Schmidt-Nielsen K, Taylor CR, Shkolnik A (1971) Desert snails: problems of heat, water and food. J Exp Biol 55:385–398

Schneider D (1966) Chemical sense communication in insects. Symp Soc Exp Biol 20:273–297

Schneider P (1971) Lebensweise und soziales Verhalten der Wüstenassel *Hemilepistus aphganicus* Borutzky 1958. Z Tierpsychol 29:121–133

Schneider P (1975) Beitrag zur Biologie der afghanischen Wüstenassel *Hemilepistus aphganicus* Borutzky 1958 (Isopoda, Oniscoidea). Aktivitätsverlauf. Zool Anz 195: 155–170

Schneirla TC (1953) Modifiability in insect behavior. In: Roeder KD (ed) Insect physiology. Wiley & Sons, New York, pp 656–779

Schone H (1984) Spatial orientation. Princeton Univ Press, Princeton

Schwartz CW, Schwartz ER (1959) The wild animals of Missouri. Univ Missouri Press, Columbia

Seely MK (1987) The Namib. Natural history of an ancient desert. Shell Oil SWA Ltd, Windhoek

Seely MK (1991) Namibia. Drought and desertification. Gamsberg Macmillan, Windhoek

Seely MK, Hamilton WJ III (1976) Fog catchment sand trenches constructed by tenebrionid beetles, *Lepidochora*, from the Namib Desert. Science 193:484–486

Selander RK (1972) Sexual selection and dimorphism in birds. In: Campbell B (ed) Sexual selection and the descent of man, 1871–1971. Aldine, Chicago, pp 180–230

Selys-Longchamps E de (1862) Catalogue raisoné des Orthoptères de Belgique. Ann Soc Entomol Belg 6:130

Serventy DL (1971) Biology of desert birds. In: Farner DS, King JR (eds) Avian biology, vol I. Academic Press, New York, pp 287–339

Serventy DL, Marshall AJ (1957) Breeding periodicity in western Australian birds: with an account of unseasonable nestings in 1953 and 1955. Emu 57:99–126

Seyfarth R, Cheney D (1990) The assessment by vervet monkeys of their own and another species' alarm calls. Anim Behav 40:754–764

Shachak M (1980) Energy allocation and life history strategy of the desert isopod *Hemilepistus reaumuri*. Oecologia 29:134–155

Shachak M, Chapman EA, Steinberger Y (1976) Feeding, energy flow and soil turnover in the desert isopod, *Hemilepistus reaumuri*. Oecologia 24:57–69

Shachak M, Steinberger Y, Orr Y (1979) Phenology, activity and regulation of radiation load in the desert isopod *Hemilepistus reaumuri*. Oecologia 40:133–140

Shaw CE (1939) Food habits of the chuckwalla, *Sauromalus obesus*. Herpetol 1:153

Shaw CE (1945) The chuckwallas, genus *Sauromalus*. Trans San Diego Soc Nat Hist 10:296–306

Shaw CE (1948) A note on the food habits of *Heloderma suspectum* Cope. Herpetol 4:145

Shaw CE (1950) Lizards in the diet of captive *Uma*. Herpetol 6:36–37

Shelly TE, Greenfield MD (1985) Alternative mating strategies in a desert grasshopper: a transitional analysis. Anim Behav 33:1211–1222

Shelly TE, Greenfield MD, Downum KR (1987) Variation in host plant quality: influences on the mating system of a desert grasshopper. Anim Behav 35:1200–1209

Shkolnik A (1971) Adaptation of animals to desert conditions. In: Evenari M, Shanon L, Tadmor N (eds) The Negev. The challenge of the deserts. Harvard Univ Press, Cambridge, pp 301–323

Shkolnik A, Borut (1969) Temperature and water relations in two species of spiny mice (*Acomys*). J Mammal 50:245–254

Shmida A, Evenari M, Noy-Meir I (1986) Hot desert ecosystems: an integrated view. In: Evenari M, Noy-Meir I, Goodall DW (eds) Hot deserts and arid shrublands B. Elsevier, Amsterdam, pp 379–387

Shoemaker VH, Nagy KA, Costa WR (1974) The consumption, utilization and modification of nutritional resources by the jackrabbit (*Lepus californicus*) in the Mojave Desert. US/IBP Desert Biome Research Memorandum, RM 74–25, Logan, Utah

Short LL (1974) Nesting of southern Sonoran birds during the summer rainy season. Condor 76:21–32

Shorthouse DJ (1971) Studies on the biology and energetics of the scorpion *Urodacus yaschenkoi*. Thesis, Aust Nation Univ, Canberra

Shulov A, Pener MP (1963) Studies on the development of eggs of the desert locust *Schistocerca gregaria* (Forsk.) and its interruption under particular conditions of humidity. Anti-Locust Bull: 41–59

Simmons JA, Brock Fenton M, O'Farrell MJ (1979) Echolocation and pursuit of prey by bats. Science 203:16–21

Simmons KEL (1951) Interspecific territorialism. Ibis 93:407–413

Simmons LW (1988) The calling song of the field cricket, *Gryllus bimaculatus* (De Geer): constraints on transmission and its role in intermale competition and female choise. Anim Behav 36:380–394

Skinner JD, Van Jaarsveld AS (1987) Adaptive significance of restricted breeding in southern African ruminants. S Afr J Sci 83:657–683
Skutch AF (1976) Parent birds and their young. Univ Texas Press, Austin
Smith SM (1975) Innate recognition of coral snake pattern by a possible avian predator. Science 187:759–760
Smith SM (1977) Coral-snake pattern recognition and stimulus generalisation by naive great kiskadees (Aves: Tyrannidae). Nature 265:535–536
Smyth M, Bartholomew GA (1966) The water economy of the black-throated sparrow and the rock wren. Condor 68:447–458
Snow DW (1961) The natural history of the oilbird, *Steatornis caripensis*, in Trinidad, W.I.: 1, general behavior and breeding habits. Zoologica 46:27–48
Soulé M (1966) Trends in the insular radiation of a lizard. Am Nat 100:47–64
Spangler HG (1984) Silence as a defense against predatory bats in two species of calling insects. Southwest Nat 29:481–488
Sperry CC (1941) Food habits of the coyote. US Fish Wildl Serv Res Bull 4:1–69
Spooner JD (1964) The Texas bush katydid. Its sounds and their significance. Anim Behav 12:235–244
Springett J (1979) (pers. comm. in, Wallwork JA 1982)
Stahnke HL (1945) Scorpions of the genus *Hadrurus* Thorell. Am Mus Novit, 1298:1–9
Stahnke HL (1966) Some aspects of scorpion behavior. Bull South Calif Acad Sci 65: 65–80
Stamps JA (1977) Social behaviour and spacing patterns in lizards. In: Gans C, Tinkle DW (eds) Biology of Reptilia 7 (Ecology and behaviour A). Academic Press, London, pp 265–334
Stebbins RC (1944) Some aspects of the ecology of the iguanid genus *Uma*. Ecol Monogr 14:311–332
Stebbins RC (1954) Amphibians and reptiles of western North America. McGraw-Hill, New York
Stemmler-Gyger O (1965) The biology of races of *Echis carinatus* (Schneider 1801). Salamandra 1:29–46
Stinner RE, Gutierrez AP, Butler GD Jr (1974) An algorithm for temperature dependent growth rate simulation. Can Entomol 106:159–524
Stone W, Rehn JAG (1903) On the terrestrial vertebrates of portions of southern New Mexico and western Texas. Proc Acad Nat Sci Phila 55:16–34
Struhsaker TT, Hunkeler P (1971) Evidence of tool-using by chimpanzees of the Ivory Coast. Folia Primatatol 15:212–219
Stuart CT (1976) Diet of the black-backed jackal *Canis mesomelas* in the central Namib Desert, South West Africa. Zool Afr 11:193–205
Sturbaum BA, Riedesel ML (1974) Temperature regulation responses of ornate box turtles, *Terrapene ornata*, to heat. Comp Biochem Physiol 48:527–528
Sublette JE, Sublette MS (1967) The limnology of playa lakes on the Llano Estacado, New Mexico and Texas. Southwest Nat 12:369–406
Sullivan BK (1989) Desert environments and the structure of anuran mating systems. J Arid Environ 17:175–183
Sutton A, Sutton M (1966) The life of the desert. Our living world of nature. McGraw-Hill, New York
Talbot L, Talbot M (1963) The wildebeest in western Masailand, East Africa. Wildl Monogr No 12
Tanner WW, Krogh JE (1974) Ecology of the leopard lizard, *Crotaphytus wislizenii*, at the Nevada test site, Nye County, Nevada. Herpetologica 30:63–72
Taylor CR (1968) Hygroscopic food: A source of water for desert antelope? Nature 219:181–182
Taylor CR (1969) The eland and the oryx. Sci Am 220:88–95
Taylor WK (1971) A breeding biology study on the verdin, *Auriparus flaviceps* in Arizona. Am Midl Nat 85:289–328

Templeton AR (1986) Coadaptation and outbreeding depression. In: Soule ME (ed) Conservation biology, Sinauer, Sunderland, pp 105–116

Tener JS (1965) Muskoxen in Canada: a biological and taxonomic review. Queen's Printer, Ottawa

Tevis L, Newell IM (1962) Studies on the biology and seasonal cycle of the giant red velvet mite, *Dinothrombium pandorae* (Acari, Trombidiidae). Ecology 43:497–505

Thomas O (1921) A new genus of opossum from southern Patagonia. Ann Mag Nat Hist 8:138

Tinbergen N (1953) Social behaviour in animals. Methuen, London

Tindale NB (1953) Tribal and intertribal marriage among Australian aborigines. Hum Biol 25:160–190

Tinkham ER (1948) Faunistic and ecological studies on the Orthoptera of the Big Bend region of Trans-Pecos Texas, with especial reference to the orthopteran zones and faunae of midwestern North America. Am Midl Nat 40:521–663

Tinkle DW (1967) The life and demography of the side-blotched lizard, *Utastansburiana*. Misc Publ Mus Zool, Univ Michigan, Ann Arbor 132:1–182

Tobias PV (1964) Bushman hunter-gatherers: a study in human ecology. In: Davis DHS (ed) Ecological studies in southern Africa. W Junk, The Hague, pp 67–78

Tomoff CS (1974) Avian species diversity in desert scrub. Ecology 55:396–403

Tongiorgi P (1962) Sulle relazioni tra habitat ed orientamento astronomico in alcune specie del gen. *Arctosa* (Araneae-Lycosidae). Boll Zool 28:683–689

Tongiorgi P (1969) Ricerche ecologiche sugli Artropodi di una spiaggia sabbiosa del litorale tirrenico. III. Migrazioni e ritmo di attività locomotoria nell'isopode *Tylos latreillei* (Aud. e Sav.) e nei tenebrionidi *Phaleria provincialis* Fauv. e *Halammobia pellucida* Herbst. Redia 51:1–19

Topoff H (1977) The pit and the antlion. Nat Hist 86:64–71

Trapp G, Hallberg DL (1975) Ecology of the gray fox (*Urocyon cinereoargenteus*): a review. In: Fox MW (ed) The wild canids. Van Nostrand Reinhold, New York, pp 164–178

Trivers RL (1971) The evolution of reciprocal altruism. Q Rev Biol 46:35–57

Tschinkel WR (1975) A comparative study of the chemical defensive system of tenebrionid beetles. Defensive behavior and ancillary features. Ann Entomol Soc Am 68:439–453

Tucker VA (1965a) Oxygen consumption, thermal conductance, and torpor in the California pocket mouse, *Perognathus californicus*. J Cell Comp Physiol 65:393–404

Tucker VA (1965b) The relation between the torpor cycle and heat exchange in the California pocket mouse *Perognathus californicus*. J Cell Comp Physiol 65:405–414

Tucker VA (1966) Diurnal torpor and its relation to food consumption and weight changes in the California pocket mouse *Perognathus californicus*. Ecology 47:245–252

Turkowski FJ, Reynolds HG (1974) Annual nutrient and energy intake of the desert cottontail, *Sylvilagus auduboni*, under natural conditions. US/IBP Desert Biome Research Memorandum, RM 74–24, Logan, Utah

Tyler MJ (1960) Observations on the diet and size variation of *Amphibolurus adelaidensis* (Gray) (Reptilia: Agamidae) on the Nullarbor Plain. Trans R Soc S Aust 83: 111–117

Ueckert DN, Bodine MC, Spears BM (1976) Population density and biomass of the desert termite *Gnathamitermes tubiformans* (Isoptera: Termitidae) in a short-grass prairie: relationship to temperature and moisture. Ecology 57:1273–1280

Uvarov BP (1966) Grasshoppers and locusts, A handbook of general acridology. I. Cambridge Univ Press, Cambridge

Uvarov BP (1977) Grasshoppers and locusts. A handbook of general acridology. II. Behaviour, ecology, biogeography, population dynamics. Centre for Overseas Pest Research, London

Van Bruggen AC (1978) Land molluscs. In: Werger MJA (ed) Biogeography and ecology of southern Africa. II. W Junk, The Hague, pp 887–923

Van der Graaf KM (1973) Comparative developmental osteology in three species of desert rodents, *Peromyscus eremicus, Perognathus intermedius* and *Dipodomys merriami.* J Mammal 54:3

Van der Merwe NJ (1953) The jackal. Fauna & Flora, Publ Transvaal Prov Admin, No 4, Pretoria

Van Lawick-Goodall H (1971) Golden jackals. In: Van Lawick-Goodall H, Van Lawick-Goodall J (eds) Innocent killers. Houghton Mifflin, Boston, pp 105–149

Van Lawick-Goodall J (1964) Tool-using and aimed throwing in a community of free-living chimpanzees. Nature 201:1264–1266

Van Lawick-Goodall J (1968) The Behavior of free-living chimpanzees in the Gombe Stream Reserve. Anim Behav Monogr 1:161–311

Van Lawick-Goodall J (1970) Tool-using in Primates and other vertebrates. In: Lehrman DS, Hinde RA, Shaw E (eds) Advances in the study of behavior. Academic Press, New York, pp 195–249

Van Lawick-Goodall J, Van Lawick-Goodall H (1966) Use of tools by the Egyptian vulture, *Neophron percnopterus*. Nature 212:1468–1469

Van Lawick-Goodall H, Van Lawick-Goodall J (1970) Innocent killers. Collins, London

Van Mierop LHS, Barnard SM (1978) Further observations on thermoregulation in the brooding female *Python molurus bivittatus* (Serpentes: Boidae). Copeia 1978: 615–621

Verhoeff KW (1935) Zur Biologie der Spirostreptiden. Zool Anz 109:288–292

Verner J (1965) Breeding biology on the long-billed marsh wren. Condor 67:6–30

Verner J, Willson MF (1969) Mating systems, sexual dimorphism, and the role of male North American passerine birds. Ornithol Monogr 9:1–76

Vince MA (1969) Embryonic communication, respiration and the synchronization of hatching. In: Hinde RA (ed) Bird vocalizations, their relation to current problems in biology and psychology. Cambridge Univ Press, London, pp 233–260

Vinciguerra MT, Zullini A (1980) New or rare species of nematodes from Italian sand dunes. Animalia 7:29–44

Vorhies CT (1928) Do southwestern quail require water? Am Nat 62:446–452

Vorhies CT, Taylor WP (1940) Life history and ecology of the white-throated woodrat, *Neotoma albigula* Hartley, in relation to grazing in Arizona. Univ Arizona Coll Agric Tech Bull 86:445–529

Wainwright CM (1978a) The floral biology and pollination ecology of two desert lupines. Bull Torr Bot Club 105:24–38

Wainwright CM (1978b) Hymenopteran territoriality and its influences on the pollination ecology of *Lupinus arizonicus*. Southwest Nat 23:605–616

Walker EP (1968) Mammals of the world, 2 vols. The Johns Hopkins Press, Baltimore

Walker TJ, Whitesell JJ (1982) Singing schedules and sites for a tropical burrowing cricket (*Anurogryllus muticus*). Biotropica 14:220–227

Wallace HR (1958) Movement of eel-worms. I. Ann Appl Biol 46:74–85

Wallace HR (1959) Movement of eelworms in water films. Ann Appl Biol 47:366–370

Wallace MMH (1968) The ecology of *Sminthurus viridis* (L.) (Collembola). II. Diapause in the aestivating egg. Aust J Zool 16:871–883

Wallace MMH (1970) Diapause in the aestivating egg of *Halotydeus destructor* (Acari: Eupodidae). Aust J Zool 18:295–313

Wallace MMH (1971) The influence of temperature and moisture on diapause development in the eggs of *Bdellodes lapidaria* (Acari: Bdellidae). J Aust Entomol Soc 10:276–280

Wallwork JA (1972) Distribution patterns and population dynamics of the microarthropods of a desert soil in southern California. J Anim Ecol 41:291–310

Wallwork JA (1976) The distribution and diversity of soil fauna. Academic Press, London

Wallwork JA (1982) Desert soil fauna. Praeger, New York

Waloff Z (1972) Orientation of flying locusts, *Schistocerca gregaria* (Forsk.), in migrating swarms. Bull Ent Res 62:1–72

Waloff Z, Rainey RC (1951) Field studies on factors affecting the displacement of desert locust swarms in eastern Africa. Anti-locust Bull 9:1–50

Walther F (1964a) Verhaltensstudien an der Gattung *Tragelaphus* De Blainville, 1816 in Gefangenschaft, unter besonderer Berücksichtigung des Sozialverhaltens. Z Tierpsychol 21:393–467, 642–646

Walther F (1964b) Einige Verhaltungsbeobachtungen an Thomson-gazellen (*Gazella thomsoni* Gunther, 1884) im Ngorongoro-Krater. Z Tierpsychol 21:871–890

Walther FR (1990) Duikers and dwarf antelopes. In: Grzimek B (ed) Grzimek's encyclopedia of mammals 5. McGraw-Hill, New York, pp 325–343

Warburg MR (1965a) On the water economy of some Australian land-snails. Proc Malacol Soc Lond 36:297–305

Warburg MR (1965b) The microclimate in the habitat of two isopods species in southern Arizona. Am Midl Nat 73:363–375

Warburg MR, Rankevich D, Chasanmus K (1978) Isopod species diversity and community structure in mesic and xeric habitats of the Mediterranean region. J Arid Environ 1:157–163

Ward D (1991) A test of the 'maxithermy' hypothesis with three species of tenebrionid beetles. J Arid Environ 21:331–336

Ward P (1979) Il mimetismo animale. (Original title: 'Colour for survival'). Ist Geogr De Agostini, Novara

Warren A (1984) The problems of desertification. In: Cloudsley-Thompson JL (ed) Sahara Desert. Pergamon Press, Oxford, pp 335–342

Washburn SL, DeVore I (1961) The social life of baboons. Sci Am 204:62–71

Watson JAL, Lendon C, Louw GN (1973) Termites in mulga lands. Tropical Grasslands 7:121–126

Watson RT, Irish J (1988) An introduction to the Lepismatidae (Thysanura: Insecta) of the Namib Desert sand dunes. Madoqua 15:285–293

Webb RG (1969) Survival adaptation of tiger salamanders (*Ambystoma tigrinum*) in the Chihuahuan Desert. In: Hoff CC, Riedesel ML (eds) Physiological systems in semiarid environments. Univ New Mexico Press, Albuquerque

Wehner R (1972) Visual orientation performance of desert ants, *Cataglyphis bicolor*, towards astronomenotactic directions and horizon landmarks. In: Galler SR, Schmidt-Koenig K, Jacobs GJ, Belleville RE (eds) Proc AIBS Symp Animal Orientation and Navigation. US Gov Print Off, Washington, pp 421–436

Wehner R, Wehner S (1986) Path integration in desert ants. Approaching a long-standing puzzle in insect navigation. Monit Zool Ital 20:309–331

Werger MJA (1986) The Karoo and southern Kalahari. In: Evenari M, Noy-Meir I, Goodall DW (eds) Hot deserts and arid shrublands, B. Elsevier, Amsterdam, pp 283–359

Werler JE (1951) Miscellaneous notes on the eggs and young of Texan and Mexican reptiles. Zoologica 36:37–48

Werner FG (1973) Foraging activity of the leaf-cutter ant, *Acromyrmex versicolor*, in relation to season, weather and colony condition. US/IBP Desert Biome Research Memorandum, RM 73–28, Logan, Utah

Werner FG, Murray SC (1972) Demography, foraging activity of leaf-cutter ants, *Acromyrmex versicolor*, in relation to colony size and location, season vegetation, and temperature. US/IBP Desert Biome Research Memorandum, RM 72–33, Logan, Utah, pp 1–10

Werner FG, Olsen AR (1973) Consumption of *Larrea* by chewing insects. US/IBP Desert Biome Research Memorandum, RM 73–32, Logan, Utah

Weygoldt P (1969) The biology of pseudoscorpions. Harvard Univ Press, Cambridge

Weygoldt P (1972) Geisselskorpione und Geisselspinnen (Uropygi und Amblypygi). Z Köln Zoo 15:95–107

Wheeler WM (1930) Demons of the dust. Norton, New York

White SR (1948) Observations on the mountain devil (*Moloch horridus*). W Aust Nat 1:78–81

Whitford WG (1973) Demography and bioenergetics of herbivorous ants in a desert ecosystem as function of vegetation, soil types and weather variables. US/IBP Desert Biome Research Memorandum, RM 73–29, Logan, Utah, pp 1–63

Whitford WG (1978a) Structure and seasonal activity of Chihuahuan desert ant communities. Ins Soc 25:79–88

Whitford WG (1978b) Foraging in seed-harvesting ants *Pogonomyrmex* spp. Ecology 59:185–189

Whitford WG, Ettershank G (1975) Factors affecting foraging in Chihuahuan desert harvester ants. Environ Entomol 4:689–696

Whitman DW, Orsak L (1985) Biology of *Taeniopoda eques* (Orthoptera: Acrididae) in southeastern Arizona. Ann Entomol Soc Am 78:811–825

Wickler W (1968) Mimetismo animale e vegetale. (Original title: 'Mimickry. Singnalfälschungen der Natur'). II Saggiatore, Milano

Wickler W (1976) The ethological analysis of attachment. Sociometric, motivational and sociophysiological aspects. Z Tierpsychol 42:12–28

Wickler W (1985) Coordination of vigilance in bird groups. The 'watchman's song' hypothesis. Z Tierpsychol 69:250–253

Wiens D (1982) Mimicry in plants. Evol Biol 11:365–403

Wiens JA (1976) Population responses to patchy environments. Annu Rev Ecol Syst 7: 81–120

Wight HM (1931) Reproduction in the eastern skink (*Mephitis mephitis nigra*). J Mammal 12:42–47

Williams CB (1958) Insect migration. Collins, London

Williams GC (1966) Adaptation and natural selection. Princeton Univ Press, Princeton

Williams SC (1966) Burrowing activities of the scorpion *Anuroctonus phaeodactylus* (Wood) (Scorpionida: Vejovidae). Proc Calif Acad Sci 34:419–428

Williams SC (1969) Birth activities of some North American scorpions. Proc Calif Acad Sci 37:1–24

Willoughby EJ (1966) Water requirements of the ground dove. Condor 68:243–248

Willoughby EJ (1971) Biology of larks (Aves: Alaudidae) in the central Namib Desert. Zool Afr 6:133–176

Wilson EO (1971) The insect societies. Belknap Press, Harvard Univ, Cambridge

Wilson EO (1975) Sociobiology, The new synthesis. Harvard Univ Press, Cambridge

Wilson RT (1989) Ecophysiology of the Camelidae and desert ruminants. Springer, Berlin Heidelberg New York

Wilson RT (1990) Natural and man-induced behaviour of the one-humped camel. J Arid Environ 19:325–340

Wood TG (1977) Food and feeding habits of termites. In: Brian MV (ed) Production ecology of ants and termites. Cambridge Univ Press, Cambridge, pp 55–80

Wood TG, Sands WA (1978) The role of termites in ecosystems. In: Brian MV (ed) Production ecology of ants and termites. IBP, vol 13. Cambridge Univ Press, Cambridge, pp 245–292

Woodbury AM (1931) A descriptive catalog of the reptiles of Utah. Bull Univ Utah 21:1–129

Woodbury AM, Hardy R (1948) Studies of the desert tortoise, *Gopherus agassizi*. Ecol Monogr 18:145–200

Woodrow DF (1965) Observations on the red locust (*Nomadacris septemfasciata* Serv.) in the Rukwa Valley, Tanganyika, during its breeding season. J Anim Ecol 34:187–200

Wooten RC Jr, Crawford CS (1974) Respiratory metabolism of the desert millipede *Orthoporus ornatus* (Girard) (Diplopoda). Oecologia 17:179–186

Wooten RC Jr, Crawford CS (1975) Food, ingestion rates, and assimilation in the desert millipede *Orthoporus ornatus* (Girard) (Diplopoda). Oecologia 20:231–236

Wright RH (1958) The olfaction guidance of flying insects. Can Entomol 90:81–89
Wylie P (1971) Cultural evolution: the fatal fallacy. Bioscience 21:729–731
Yeates GW (1967) Studies on nematodes from dune sands. 9. Quantitative comparison of the nematode faunas of six localities. N Z J Sci 10:927–948
Yom-Tov Y (1970) The effect of predation on population densities of some desert snails. Ecology 51:907–911
Yom-Tov Y (1971) The biology of two desert snails *Trochoidea (Xerocrassa) seetzeni* and *Sphincterochila boissieri*. Isr J Zool 20:231–248
Yom-Tov Y (1972) Field experiments on the effect of population density and slope direction on the reproduction of the desert snail *Trochoidea (Xerocrassa) seetzeni*. J Anim Ecol 41:17–22
Young G (1962) The hill tribes of northern Thailand: a socio-ethnological report, 2nd edn. Siam Soc, Bangkok
Young SP (1946) The puma-mysterious American cat. Dover, New York
Young SP (1958) The bobcat of North America. Wildl Manag Inst, Washington, DC
Young SP (1964) The wolves of North America. Part I. Their history, life habits, economic status and control. Dover, New York
Youthed GJ, Moran VC (1969) The lunar-day activity rhythm of Myrmeleontid larvae. J Insect Physiol 15:1259–1271
Zahavi A (1974a) Altruism in babblers as a communication system. Abstr, Int Ethol Conf, Parma, Italy
Zahavi A (1974b) Communal nesting by the Arabian babbler: a case of individual selection. Ibis 116:84–87
Zann R (1977) Pair-bond and bonding behaviour in three species of grassfinches of the genus *Poephila* (Gould). Emu 77:97–106
Zannier F (1965) Verhaltensuntersuchungen an der Zwergmanguste, *Helogale undulata rufula* im Zoologischen Garten Frankfurt am Main. Z Tierpsychol 22:672–695
Zweifel RG (1968) Reproductive biology of anurans of the arid southwest, with emphasis on adaptation of embryos to temperature. Bull Am Mus Nat Hist 140:1–64

Subject Index

Springer-Verlag and the Environment

We at Springer-Verlag firmly believe that an international science publisher has a special obligation to the environment, and our corporate policies consistently reflect this conviction.

We also expect our business partners – paper mills, printers, packaging manufacturers, etc. – to commit themselves to using environmentally friendly materials and production processes.

The paper in this book is made from low- or no-chlorine pulp and is acid free, in conformance with international standards for paper permanency.

Printing: Saladruck, Berlin
Binding: Buchbinderei Lüderitz & Bauer, Berlin